Lecture Notes in Physics

The Editorial Policy for Proceedings

The series Lecture Notes in Physics reports new developments in physical research and teaching – quickly, informally, and at a high level. The proceedings to be considered for publication in this series should be limited to only a few areas of research, and these should be closely related to each other. The contributions should be of a high standard and should avoid lengthy redraftings of papers already published or about to be published elsewhere. As a whole, the proceedings should aim for a balanced presentation of the theme of the conference including a description of the techniques used and enough motivation for a broad readership. It should not be assumed that the published proceedings must reflect the conference in its entirety. (A listing or abstracts of papers presented at the meeting but not included in the proceedings could be added as an appendix.)

When applying for publication in the series Lecture Notes in Physics the volume's editor(s) should submit sufficient material to enable the series editors and their referees to make a fairly accurate evaluation (e.g. a complete list of speakers and titles of papers to be presented and abstracts). If, based on this information, the proceedings are (tentatively) accepted, the volume's editor(s), whose name(s) will appear on the title pages, should select the papers suitable for publication and have them refereed (as for a journal) when appropriate. As a rule discussions will not be accepted. The series editors and Springer-Verlag will normally not interfere with the detailed editing except in fairly obvious cases or on technical matters.

Final acceptance is expressed by the series editor in charge, in consultation with Springer-Verlag only after receiving the complete manuscript. It might help to send a copy of the authors' manuscripts in advance to the editor in charge to discuss possible revisions with him. As a general rule, the series editor will confirm his tentative acceptance if the final manuscript corresponds to the original concept discussed, if the quality of the contribution meets the requirements of the series, and if the final size of the manuscript does not greatly exceed the number of pages originally agreed upon. The manuscript should be forwarded to Springer-Verlag shortly after the meeting. In cases of extreme delay (more than six months after the conference) the series editors will check once more the timeliness of the papers. Therefore, the volume's editor(s) should establish strict deadlines, or collect the articles during the conference and have them revised on the spot. If a delay is unavoidable, one should encourage the authors to update their contributions if appropriate. The editors of proceedings are strongly advised to inform contributors about these points at an early stage.

The final manuscript should contain a table of contents and an informative introduction accessible also to readers not particularly familiar with the topic of the conference. The contributions should be in English. The volume's editor(s) should check the contributions for the correct use of language. At Springer-Verlag only the prefaces will be checked by a copy-editor for language and style. Grave linguistic or technical shortcomings may lead to the rejection of contributions by the series editors. A conference report should not exceed a total of 500 pages. Keeping the size within this bound should be achieved by a stricter selection of articles and not by imposing an upper limit to the length of the individual papers. Editors receive jointly 30 complimentary copies of their book. They are entitled to purchase further copies of their book at a reduced rate. As a rule no reprints of individual contributions can be supplied. No royalty is paid on Lecture Notes in Physics volumes. Commitment to publish is made by letter of interest rather than by signing a formal contract. Springer-Verlag secures the copyright for each volume.

The Production Process

The books are hardbound, and the publisher will select quality paper appropriate to the needs of the author(s). Publication time is about ten weeks. More than twenty years of experience guarantee authors the best possible service. To reach the goal of rapid publication at a low price the technique of photographic reproduction from a camera-ready manuscript was chosen. This process shifts the main responsibility for the technical quality considerably from the publisher to the authors. We therefore urge all authors and editors of proceedings to observe very carefully the essentials for the preparation of camera-ready manuscripts, which we will supply on request. This applies especially to the quality of figures and halftones submitted for publication. In addition, it might be useful to look at some of the volumes already published. As a special service, we offer free of charge LATEX and TEX macro packages to format the text according to Springer-Verlag's quality requirements. We strongly recommend that you make use of this offer, since the result will be a book of considerably improved technical quality. To avoid mistakes and time-consuming correspondence during the production period the conference editors should request special instructions from the publisher well before the beginning of the conference. Manuscripts not meeting the technical standard of the series will have to be returned for improvement.

For further information please contact Springer-Verlag, Physics Editorial Department V, Tiergartenstrasse 17, D-69121 Heidelberg, FRG

J. L. Sanz   E. Martínez-González   L. Cayón   (Eds.)

Present and Future of the Cosmic Microwave Background

Proceedings of the Workshop
Held in Santander, Spain
28 June - 1 July 1993

Springer-Verlag
Berlin Heidelberg GmbH

Editors

José Luis Sanz
Enrique Martínez-González
Laura Cayón
Departamento de Física Moderna, Facultad de Ciencias
Universidad de Cantabria, Avda. Los Castros s/n
E-39005 Santander (Cantabria), Spain

Local Organizing Committee

J. L. Sanz, E. Martínez-González and L. Cayón
Universidad de Cantabria, Spain

International Organizing Committee

E. Bertschinger, R. Davies, B. J. T. Jones, F. Melchiorri, J. Silk, G. Smoot

ISBN 978-3-662-13992-9   ISBN 978-3-540-48328-1 (eBook)
DOI 10.1007/978-3-540-48328-1

© Springer-Verlag Berlin Heidelberg 1994
Originally published by Springer-Verlag Berlin Heidelberg New York in 1994
Softcover reprint of the hardcover 1st edition 1994

This book was processed using the LᴬTEX macro package with LMAMULT style

SPIN: 10080329   58/3140-543210 - Printed on acid-free paper

Preface

The Workshop *"Present and Future of the Cosmic Microwave Background"* was held in Santander (Spain), June 28 - July 1, 1993, at the Universidad Internacional Menéndez Pelayo (U.I.M.P.).

The idea was to review and discuss the most recent developments in this field as well as the future prospects. The present status of the observations of the spectrum and anisotropies of the *cosmic microwave background* (CMB) were presented by invited speakers. The Workshop also intended to cover experimental developments, data analysis and theoretical aspects related to this background.

We had also in mind the idea of promoting scientific collaborations and contacts at the European level, in fact many people came from the different laboratories that are now collaborating in the European Network on the CMB (Santander, Tenerife, Manchester, Oxford, Rome and Paris).

The last decade has been very successful for cosmology. On the theoretical side, the *inflationary* model has originated a paradigm giving a global density parameter $\Omega \simeq 1$ and the primordial spectrum of the density perturbations. On the observational side, the emergence of *large-scale structure* (big voids, the great wall,...) in the universe is a real fact, but the most relevant contribution –if confirmed– is without any doubt the one by COBE. The FIRAS instrument has confirmed the prediction of a *black-body* spectrum for the cosmic microwave background (CMB) over a wide range covering the submillimeter region and this is a strong support for the big-bang model, whereas the DMR experiment has detected anisotropy in the CMB at the level 10^{-5} on angular scales above $10°$. This level of anisotropy is consistent with the inflationary scenario based on a scale-invariant spectrum and, to a certain extent, confirms that our ideas about gravitational instability operating on initial seeds to form galaxies, clusters, etc. are along the right lines.

These proceedings contain the review talks and contributions presented at the workshop.

The organizers express their cordial thanks to all participants, and especially to our speakers who kindly accepted our invitation. We are also indebted to the sponsoring institutions: U.I.M.P. and Universidad de Cantabria (STRIDE Programme of the EEC) and as a collaborator Facultad de Ciencias de la Universidad de Cantabria.

Santander
October 1993

J. L. Sanz
E. Martínez-González
L. Cayón

Contents

The CMB Spectrum at Centimeter Wavelengths

$M.Bersanelli^1$, $G.F.Smoot^2$, $M.Bensadoun^2$, $G.De\ Amici^2$ and $M.Limon^2$

[1] Istituto di Fisica Cosmica, CNR, 20133 Milano, Italy
[2] Lawrence Berkeley and Space Science Laboratory, Berkeley, CA 94720, USA

ABSTRACT - The results of ground-based measurements of the cosmic microwave background (CMB) spectrum at cm-wavelengths are discussed. We report on the analysis of our most recent measurement at a frequency of 2 GHz (15 cm wavelength) in the context of the present observational situation.

1 Introduction

The low-frequency portion of the CMB spectrum is expected to exhibit the largest deviations from a purely planckian distribution in the event of energy releases in the early ($z \lesssim 3 \times 10^6$) Universe. Theoretical predictions of spectral distortions have been investigated soon after the CMB discovery [1,2] and studied in greater detail in recent works (e.g. [3,4] and references therein). Since the early 80's an Italian-American collaboration has performed several ground-based absolute measurements of the CMB spectrum in the Rayleigh-Jeans region [5,6,7] progressively improving the observational limits and extending the frequency coverage. The measurements from 1982 to 1988 were performed in 6 campaigns from the White Mountain Research Station, California, while the last two sets of measurements were taken from the South Pole. Fig. 1 describes the experiment technique used above 1 GHz. Each radiometer measures the signal difference, ΔS, between the zenith sky and a calibrating blackbody source cooled at liquid helium temperature whose antenna temperature[3], $T_{A.load}$, is precisely known. To derive the CMB antenna temperature, $T_{A.CMB}$, all the *local* contributions to the zenith sky signal need to be evaluated: at centimeter wavelengths they are dominated by the emission from the atmosphere, $T_{A.atm}$, the Galaxy, $T_{A.Gal}$, and the ground, $T_{A.ground}$:

$$T_{A.CMB} = G(\Delta S) + T_{A.load} - \delta T_{inst} - T_{A.atm} - T_{A.Gal} - T_{A.ground}$$

The radiometer calibration constant, G, is repeatedly measured during the experiment. The term δT_{inst} refers to changes in the radiometer performance due to the inversion of the instrument during the calibration. Generally, the accuracy of the measurement is limited by the systematic uncertainties related to the subtracted foreground components.

---

[3] The antenna temperature is defined as $T_A = P/kB = T_\nu [exp(T_\nu/T) - 1]^{-1}$, where $T_\nu = h\nu/k$, P is the power intercepted by the antenna, and B is the bandwidth.

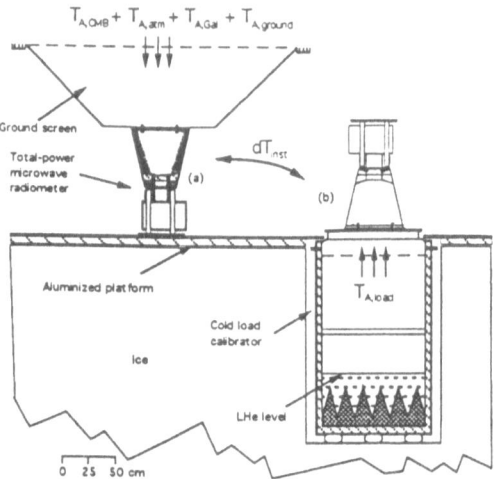

$$T_{A,CMB} + T_{A,atm} + T_{A,Gal} + T_{A,ground}$$

Ground screen

Total-power microwave radiometer

(a)

dT_{inst}

(b)

Aluminized platform

Cold load calibrator

Ice

LHe level

$T_{A,load}$

0 25 50 cm

Fig. 1. Schematic of the measurement technique for the 2 GHz radiometer (South Pole). The concept applies to other cm-wavelength measurements.

2 The Measurement at 2 GHz

We designed our new instrument to measure at a frequency of 2 GHz, where significant distortions can be present and the Galactic foreground is still an order of magnitude lower than the CMB signal at high Galactic latitudes. The 2 GHz radiometer used a rectangular, E-plane corrugated horn, and a total power, RF-gain receiver with a low-loss front-end filter [8,9].

Even from a dry, high-altitude site as the South Pole the emission from the atmosphere is the largest correction at 2 GHz, being $\sim 40\%$ of the CMB signal. We directly measured $T_{A.atm}$ with the 2 GHz radiometer by measuring the differential emission at zenith angles $0°\text{-}30°$, $0°\text{-}40°$, $0°\text{-}50°$. We observed sky regions with small (< 0.1 K) differential Galactic signal (RA$\sim 5^h$) to minimize the error due to the related correction. Including systematic uncertainties we find $T_{A.atm} = 1.04 \pm 0.10$ K. We also obtain an independent evaluation of $T_{A.atm}$ by extrapolating to 2 GHz our measurements at 3.8 GHz and 7.5 GHz from the same site. The high frequency measured values are corrected for the effect of the different beam pattern and fitted to the spectral shape predicted by models of atmospheric emission. We find $T_{A.atm} = 1.08 \pm 0.07$ K, in good agreement with the measured value.

The emission from the ground and from the Sun was minimized by the design of the antenna and by shielding the instrument with large aluminum reflectors, both during absolute and differential measurements. We evaluate the effect of ground emission (~ 50 mK level) with simulations, which yield results consistent with lower limits placed by specific tests performed at the site.

To subtract the Galactic emission we rely on existing low-frequency maps [10] and evaluations of the spectral index [11]. We convolve the high resolution (0.85°) 408 MHz Haslam map to our antenna beam pattern (HPBW$\simeq 22°$) after correcting for HII Galactic emission. In fig. 2a we show a histogram of all our measurements of $T_{A.sky} \equiv T_{A.CMB} + T_{A.Gal}$, i.e., after all foregrounds except the Galaxy have been removed. As a crosscheck, one out of the six runs of absolute calibration (dark area) was performed pointing the antenna at $\delta = -74°$, RA= $3^h\ 5^m$, a direction where the Galactic emission was $\sim 25\%$ lower than at Zenith ($\delta = -90°$). Fig. 2b shows the histogram for $T_{A.CMB}$, i.e., after $T_{A.Gal}$ has been subtracted from each run. When we convert $T_{A.CMB}$ into thermodynamic temperature we find $T_{CMB}(2\ \mathrm{GHz}) = 2.55 \pm 0.15$ K, where the errorbar is 68% confidence level and dominated by systematics.

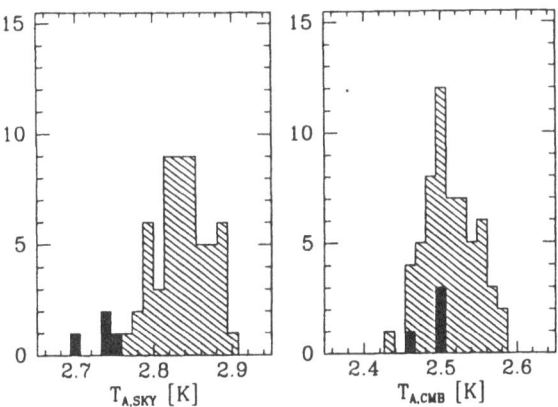

Fig. 2. Histograms of the sky antenna temperature (left) and CMB antenna temperature (right). Dark area represent data from run n.6.

3 Overall Ground-Based Results

The 2 GHz measurement is the latest achievement of a larger collaborative effort (Table 2; [5,6]) to characterize the centimeter range of the CMB spectrum. Measurements were obtained at 13 different wavelengths spanning over two decades in frequency. The best fit blackbody spectrum to ground-based measurements gives $T_{CMB} = 2.64 \pm 0.04$ K, or about 80 mK lower than the average results at higher frequencies [12,13]. We have been aware of this apparent discrepancy since high frequency measurements, such as those based on interstellar CN, have become sufficiently accurate[4]. We repeated measurements at constant frequencies with improvements and changes in the hardware and from different sites, to search for possible undetected overall systematic errors. However, we have always found self-consistent results, and all the measurements performed from both White Mountain and the South Pole agree within 1σ (see Table 1).

---

[4] It should be noted that CN-measurements now show an excess of 80 ± 32 mK over the FIRAS and COBRA results (see [14] for a discussion).

Table 1

References	ν GHz	λ cm	Campaign[a]	$T_{A.atm}$ K	$T_{A.Gal}$ K	T_{CMB} K
Sironi et al. 1990, ApJ, 357, 301	0.60	50	AG1986	1.170±0.300	7.010±0.870	3.00±1.20
Sironi et al. 1991, ApJ, 378, 550	0.82	36	SP1989	0.900±0.350	3.010±0.340	2.70±1.60
Levin et al. 1988, ApJ, 334, 14	1.41	21.3	WM1986	0.830±0.100	0.800±0.160	2.11±0.38
Bensadoun et al. 1993, ApJ, 409, 1	1.47	20.4	WM1988	0.935±0.070	1.498±0.317	2.27±0.25
			SP1989	1.046±0.076	0.819±0.205	2.26±0.21
			SP1991[b]	...	...	...
Bersanelli et al. 1993, ApJ, in press	2.0	15	SP1991	1.065±0.070	0.330±0.098	2.55±0.14
Sironi & Bonelli 1986, ApJ, 311, 418	2.5	12	WM1982	0.950±0.050	0.148±0.030	2.62±0.25
Sironi et al. 1991, ApJ, 378, 550			WM1983	0.950±0.050	0.200±0.030	2.79±0.15
			SP1989	1.155±0.300	0.134±0.025	2.50±0.34
De Amici et al. 1988, ApJ, 329, 556	3.7	8.1	WM1986	0.870±0.108	0.066±0.030	2.59±0.13
De Amici et al. 1991, ApJ, 381, 341	3.8	7.9	WM1987	0.898±0.064	0.057±0.010	2.56±0.08
			WM1988	0.955±0.055	0.060±0.010	2.71±0.07
			SP1989	1.109±0.060	0.055±0.015	2.64±0.07
Mandolesi et al. 1986, ApJ, 310, 561	4.75	6.3	WM1982	1.000±0.100	0.040±0.030	2.73±0.22
			WM1983	0.997±0.070	0.035±0.025	2.70±0.07
Kogut et al. 1990, ApJ, 355, 102	7.5	4.0	WM1988	1.175±0.078	0.010±0.005	2.60±0.07
Levin et al. 1992, ApJ, 396, 3			SP1989	1.222±0.059	0.007±0.004	2.69±0.07
			SP1991[b]	...	...	...
Kogut et al. 1988, ApJ, 235, 1	10	3.0	WM1982	1.190±0.113	0.003±0.003	2.91±0.17
			WM1983	1.200±0.130	0.004±0.002	2.64±0.14
			WM1984	1.122±0.120	0.004±0.002	2.65±0.21
			WM1986	1.222±0.065	0.008±0.004	2.56±0.08
			WM1987	1.173±0.086	0.006±0.003	2.62±0.09
De Amici et al. 1985, ApJ, 298, 710	33	0.9	WM1982	4.850±0.140	0.001±0.001	2.82±0.21
			WM1983	4.530±0.090	0.001±0.001	2.81±0.14
			WM1984	4.340±0.090	0.001±0.001	2.81±0.14
Witebsky et al 1986, ApJ, 310, 145	90	0.3	WM1982	12.600±0.570	0.001±0.001	2.58±0.74
Bersanelli et al. 1989, ApJ, 339, 632			WM1983	9.870±0.090	0.001±0.001	2.57±0.12
			WM1984	11.300±0.130	0.001±0.001	2.53±0.18
			WM1986	15.020±0.100	0.001±0.001	2.68±0.14
			WM1987	13.840±0.035	0.001±0.001	2.60±0.11
			WM1988	9.360±0.040	0.001±0.001	...
			WM1989	8.800±0.020	0.001±0.001	...

[a] AG: Alpe Gera, Italy; WM: White Mountain, USA; SP: South Pole, Antarctica.
[b] Analysis in progress.

An obvious candidate for an overall systematic bias is an underestimate of $T_{A.load}$, since the cold load calibrator [15] is a piece of equipment shared by different radiometers. Note however that before 1988 another cold load was used, similar in design but with different corrections to be applied to the liquid helium boiling temperature. Recently we directly tested the cold load to measure the emission from the internal radiometric walls of the dewar and found no measurable effect (< 40 mK upper limit) at 2 GHz. A systematic overestimate of $T_{A.atm}$ could also produce the observed discrepancy. However one would have to explain the internal consistency of our atmospheric data set. We find very good agreement in all measurements (from 2 to 90 GHz) between evaluations of $T_{A.atm}$ based on different scan angles; our results fit well the spectral shape expected from atmospheric models; finally, we find consistency in our $T_{A.CMB}$ results obtained from sites with significantly different atmospheric emission. The foreground correction with the highest relative uncertainty is the Galactic emission. The uncertainties in the 408 MHz map and in α_{syn} dominate the error on T_{CMB} below 2.5 GHz. However, at frequencies $\gtrsim 6$ GHz the Galactic emission is small enough that *any* overestimate of $T_{A.Gal}$ would not significantly affect the results. We have so far been unable to detect overall systematic errors throughout our measurements. It is highly unlikely that a single source of error can fully reconcile the low and high frequency data, although it is conceivable that a conspiracy could do that.

4 Conclusions

At present ground-based results provide the best observational limits to the CMB spectrum at centimeter wavelengths. They can be used in conjunction with other measurements to constrain models of expected spectral distortions. In fig. 4 we show the maximum μ-distortion allowed by FIRAS and by using the low-frequency data: it seems unlikely that future progress in low-frequency spectral measurements may improve limits on such distortion models. On the other hand free-free distortions (as can be expected from re-ionization processes or non-recombination models) are significantly constrained by cm-wavelength results. Using all the available measurements we find a 2σ upper limit to the free-free parameter $Y_{ff} \equiv \int (1 - T_c/T_\gamma)\kappa dt < 1.9 \times 10^{-5}$. The best fit suggests a negative free-free parameter ($Y_{ff} = -6.5 \pm 8.4 \times 10^{-5}$, 2σ) which would imply an electron temperature, T_e, lower than radiation temperature, T_γ.

Accurate measurements of the CMB spectrum at cm-wavelengths require significant progress in our understanding of the Galactic emission. An improvement by a factor of 3 in the determination of α_{syn} above 408 MHz and by a factor of 2 in the absolute calibration of the Haslam map would greatly enhance the quality of our results. The same data obtained in past campaigns (Table 2) could be reanalyzed using the new measured Galactic parameters and one can expect to constrain free-free distortions over an order of magnitude better. Such progress is within the reach of present technology and require relatively inexpensive, though long-term projects. New collaborative efforts between groups from

Berkeley, Milano, Rome, and South American Institutions are underway to produce absolutely calibrated maps at several frequencies between 0.4 and 5 GHz. In our 1991 South Pole campaign we performed a measurement at 408 MHz using a prototype instrument to scan the sky at $\delta = -60°$, and gained experience for future measurements [16]. Improved instruments are now under construction by the Berkeley and Milano groups. This project is also expected to be extremely beneficial to present and future measurements of the CMB anisotropy which are now reaching sensitivity levels $\Delta T/T \sim 10^{-5}$–10^{-6}, i.e., the level expected for Galactic foreground confusion and CMB anisotropy detection.

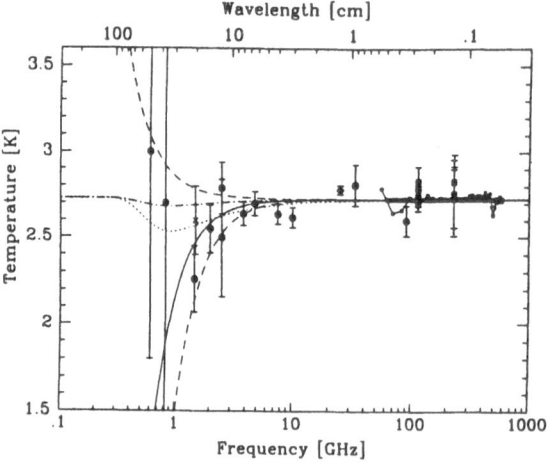

Fig. 3. Recent measurements of the CMB spectrum and distortion models: *solid line -* Best fit free-free distortion; *dashed -* 2σ limits to free-free; *dot-dashed -* FIRAS limit to μ-distortions; *dotted -* Ground-based limit to μ-distortions. Filled circles are results from the Italy-USA collaboration.

References

1. Peebles, P.J.E. 1968, ApJ, 153, 1
2. Zel'dovich, Ya. B., Kurt, V.G., Sunyaev, R.A. 1969, Sov. Phys. JETP, 28, 146
3. Silk, J. & Stebbins, A. 1983, ApJ, 269, 1
4. Burigana, C., Danese, L. & De Zotti, G. 1991 A&A, 246, 49
5. Smoot, G.F. et al. 1983, Phys. Rev. Lett., 51, 1099
6. Smoot, G.F. et al. 1987, ApJ, 317, L45
7. Sironi, G. et al. 1990, ApJ, 357, 301
8. Bersanelli et al. 1993, ApJ, in press
9. Bersanelli et al. 1992, IEEE Trans. Antennas Propagat., 40, 1107
10. Haslam, C.G.T et al. 1982, A&A Suppl., 47, 1
11. Lawson, K.D. et al. 1987, MNRAS, 225, 307
12. Mather, J. et al. 1993, ApJ Lett, in press
13. Gush, H.P., Halpern, M., & Wishnow, E.H. 1990, Phys. Rev. Lett., 65, 537
14. Palazzi, E., Mandolesi, N. & Crane, P. 1992, ApJ, 398, 53
15. Bensadoun, M. et al. 1992, Rev. Sci. instrum. 63, 4377
16. De Amici, G. et al. 1993, in *Observational Cosmology*, Chincarini et al. ed., ASP vol. 51, p.527

Recent Measurements of the Sunyaev-Zel'dovich Effect

Mark Birkinshaw

Smithsonian Astrophysical Observatory, 60 Garden Street, Cambridge, MA 02138, USA

1 Abstract

Several techniques for the measurement of the Sunyaev-Zel'dovich effects in clusters of galaxies are now yielding reliable results. The data are being used to study cluster atmospheres, measure the Hubble constant, and search for cluster peculiar velocities. This review summarizes the observational status of the Sunyaev-Zel'dovich effects and the implications of recent results.

2 The Effects

The Sunyaev-Zel'dovich effects (Sunyaev & Zel'dovich 1972, 1980) arise from inverse-Compton scatterings of photons of the microwave background radiation (which has temperature $T_r = 2.74$ K, Mather *et al.* 1990) by electrons in a gas at temperature $T_e \gg T_r$. On average, an inverse-Compton scattering causes a photon's energy to increase by an amount proportional to $k_B T_e/m_e c^2$, and the optical depth to such scatterings is $\tau_e \approx n_e \sigma_T d$, where k_B is the Boltzmann constant, m_e is the electron rest mass, c is the speed of light, n_e is the electron concentration, σ_T is the Thomson scattering cross-section, and d is the path length through the scattering medium. The fractional change in the specific intensity, I_ν, of the background radiation as viewed through the scattering medium is proportional to the product of these terms, and is a decrease

$$\frac{\Delta I_\nu}{I_\nu} \approx -2\ \tau_e\ \frac{k_B T_e}{m_e c^2} \tag{1}$$

in the Rayleigh-Jeans part of the spectrum. A second effect,

$$\frac{\Delta I_\nu}{I_\nu} \approx -\tau_e\ \frac{v_r}{c}\ , \tag{2}$$

arises if the scattering medium is in motion (with peculiar radial velocity v_r) relative to the reference frame defined by the background radiation.

The largest detectable Sunyaev-Zel'dovich effects are expected from clusters of galaxies, for which X-ray data have demonstrated the presence of a hot

$(k_\mathrm{B}T_\mathrm{e} \approx 8$ keV$)$ and relatively dense $(n_\mathrm{e} \approx 2 \times 10^3$ m$^{-3})$ intracluster medium with a scale $d \approx 1$ Mpc. Such gas has $\tau_\mathrm{e} \approx 0.004$, and causes a fractional energy gain ≈ 0.015 per scattering, so that the thermal Sunyaev-Zel'dovich effect is expected to be $\Delta I_\nu / I_\nu \approx -8 \times 10^{-5}$, corresponding to a brightness temperature change $\Delta T_\mathrm{RJ} \approx -0.3$ mK. The kinematic effect is smaller by a factor $\approx 0.11\,(v_\mathrm{r}/1000$ km s$^{-1})\,(k_\mathrm{B}T_\mathrm{e}/8$ keV$)^{-1}$, relatively small for hot clusters. The flux density of the Sunyaev-Zel'dovich effect from the core of a cluster of galaxies at wavelength λ is then $\Delta S_\mathrm{core} \approx 5\,(\Delta T_\mathrm{RJ}/\mathrm{mK})\,(\lambda/\mathrm{cm})^{-2}\,(\theta_\mathrm{core}/\mathrm{arcmin})^2$ mJy, where θ_core is the X-ray core radius of the cluster.

Although the brightness temperature effects are almost frequency independent at $\nu \lesssim 50$ GHz, outside the Rayleigh-Jeans regime they have different spectra (Fig. 1). The largest thermal and kinematic effects are both seen at zero frequency, but the thermal effect changes sign at 220 GHz and reaches a positive peak at 310 GHz while the kinematic effect remains negative (for positive v_r). Accurate spectral measurements of the combined effect towards a cluster of galaxies should be capable of separating the thermal and kinematic components: e.g., observations at 220 GHz measure only the kinematic term $(\Delta T_\mathrm{RJ}(220$ GHz$) = 0.33\Delta T_\mathrm{K0})$.

The amplitudes of the Sunyaev-Zel'dovich effects $(\Delta T_\mathrm{T0}$ and $\Delta T_\mathrm{K0})$ depend only on the physical properties of the cluster producing them. Clusters with the same properties at different redshifts therefore display the same brightness temperature effects. This distance independence of the thermal Sunyaev-Zel'dovich effect makes it a sensitive probe of distant clusters and their evolution (e.g., Markevitch *et al.* 1992, 1993; Bartlett & Silk 1993).

Inverse-Compton scatterings are a feature of non-thermal plasmas as well as thermal plasmas, so that a Sunyaev-Zel'dovich effect may also be expected from the radio-emitting plasma in the diffuse lobes of a radio source (McKinnon *et al.* 1990), although it may be difficult to detect near bright radio emission. No detections of this effect have been reported.

3 Techniques

Three distinct techniques are in use for measuring the Sunyaev-Zel'dovich effects of clusters: single-dish radiometry, bolometric observations, and interferometry. Table 1 lists measurements of cluster effects over the past 10 years and gives the significance of the best-detected effect in each paper.

3.1 Single-dish radiometry

The most heavily used technique, to date, is that of single-dish radiometry, where the brightness of the radio sky towards a cluster of galaxies is measured using a radiometer mounted on a large radio telescope.

In order to reduce the effects of the atmosphere above the telescope a differential (typically twin-beam) system is used, and the data record the difference in the brightnesses of matched beams (of full-width to half-maximum θ_h) separated on the sky by an angle $\theta_\mathrm{sw} > \theta_\mathrm{h}$. A variety of switching schemes have

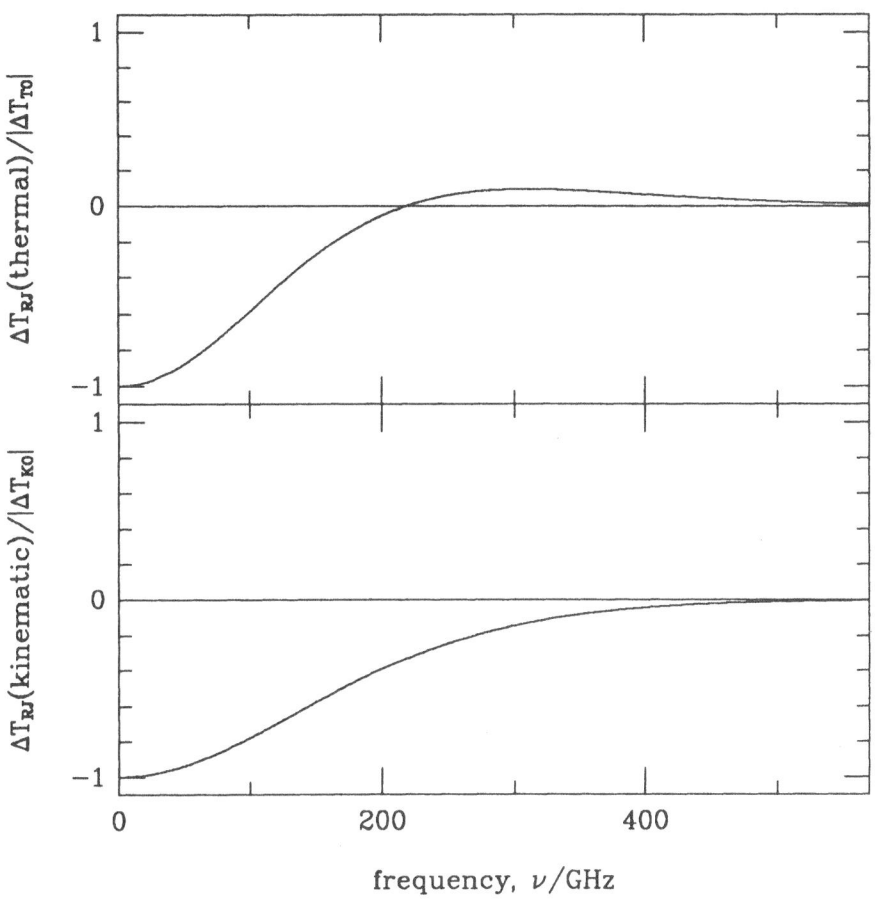

Fig. 1. The spectra of the thermal and kinematic Sunyaev-Zel'dovich effects. Their amplitudes, ΔT_{T0} and ΔT_{K0}, may both be measured from accurate spectral data.

been used in attempts to optimize the removal of atmospheric signals (which are $> 10^3$ times brighter than the Sunyaev-Zel'dovich effects): these schemes are described in detail in the published work (e.g., Birkinshaw & Gull 1984). Although beam-switching is efficient at subtracting atmospheric effects, it restricts the choice of clusters. Clearly the technique fails for clusters at low redshift that have angular sizes $\gg \theta_{sw}$, because the 'off' beam positions are contaminated by the Sunyaev-Zel'dovich effect. (The angular size of a cluster in the Sunyaev-Zel'dovich effect is a factor $2 - 4$ larger than in the X-ray surface brightness). The technique is limited at high redshifts by the effects of beam dilution when observing clusters with angular size $\ll \theta_h$, but since the angular size of a cluster changes only slowly with redshift at $z \gtrsim 0.5$, this limitation is often not severe. Figure 2 shows the variation with redshift of the observable central Sunyaev-Zel'dovich effect from a typical cluster for observations with the OVRO 40-m

Table 1. The Sunyaev-Zel'dovich effect in the last decade

reference	detection significance
Single-dish radiometers	
Andernach *et al.* 1983	1σ
Lasenby & Davies 1983	0σ
Birkinshaw & Gull 1984	4σ
Birkinshaw *et al.* 1984	7σ
Uson 1985	3σ
Andernach *et al.* 1986	4σ
Uson 1987	2σ
Klein *et al.* 1991	3σ
Birkinshaw *et al.* 1993	7σ
Herbig *et al.* 1993	5σ
Bolometers	
Meyer *et al.* 1983	0σ
Radford *et al.* 1986	1σ
Chase *et al.* 1987	2σ
Wilbanks *et al.* 1993	6σ
Interferometers	
Partridge *et al.* 1987	0σ
Jones *et al.* 1993	5σ

telescope at 20 GHz ($\theta_h = 1.8$ arcmin, $\theta_{sw} = 7.1$ arcmin).

Another difficulty is that much of this work is done at cm-wavelengths, where large antennas are readily available, and the atmosphere is relatively benign, but where the radio sky is confused by non-thermal sources associated with galaxies (in the target cluster, the foreground, or the background) and quasars. The effects of these radio sources must be subtracted if the Sunyaev-Zel'dovich effects are to be seen cleanly.

Herbig *et al.* (1993) have recently detected the Sunyaev-Zel'dovich effect of the Coma cluster using these methods on the OVRO 5.5-m telescope at 32 GHz. At this frequency the telescope provides beams with $\theta_h = 7$ arcmin separated by $\theta_{sw} = 22$ arcmin, and a three-stage differencing scheme was used to eliminate atmospheric and other error signals. Their result, an antenna temperature effect of -175 ± 21 μK, corresponding to a central Sunyaev-Zel'dovich effect $\Delta T_{RJ0}(= \Delta T_{T0} + \Delta T_{K0}) = -510 \pm 110$ μK, is a convincing measurement of the Sunyaev-Zel'dovich effect from a nearby, well-studied, cluster of galaxies.

More distant clusters have been the subject of recent work by Birkinshaw *et al.* (1993), who used the OVRO 40-m telescope at 20 GHz to measure the amplitudes and angular structures of the Sunyaev-Zel'dovich effects of 0016+16, Abell 665, and Abell 2218. The scan data for these clusters are shown in Fig. 3: the centers of the Sunyaev-Zel'dovich effects are consistent with the X-ray centers of the clusters, and the angular structures are consistent with simple models of

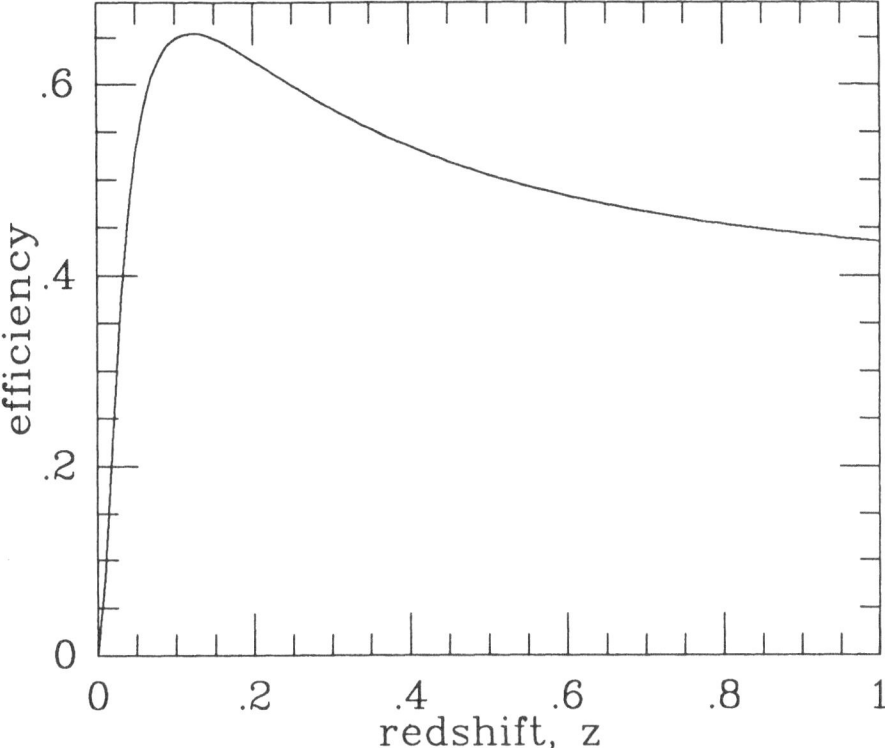

Fig. 2. The redshift dependence of the observing efficiency of the OVRO 40-m telescope at 20 GHz for clusters with core radius 300 kpc. The efficiency, η, is the central effect seen by the telescope divided by the true amplitude of the Sunyaev-Zel'dovich effect, and measures the beam-dilution and beam-switching reductions of the cluster signal. The decrease in η at $z > 0.15$ is slow, so that the 40-m telescope is sensitive to the Sunyaev-Zel'dovich effects of clusters over a wide redshift range.

the cluster atmospheres. The errors vary significantly over the three scans, partly because of different corrections for radio source contamination (several points are near sources brighter than $100\,\mu$K, and several of the sources are variable: Moffet & Birkinshaw 1989), but also because the errors include estimates of position-dependent systematic errors.

3.2 Bolometric methods

The principal advantage of a bolometric system is the high sensitivity that is achieved, but these devices are also of interest because of their frequency range: at present they provide the best sensitivity for observing the microwave background outside the Rayleigh-Jeans part of the spectrum, and hence for detecting the kinematic component of the Sunyaev-Zel'dovich effect. Furthermore, the best systems consist of several detectors arranged in an array, and some provide si-

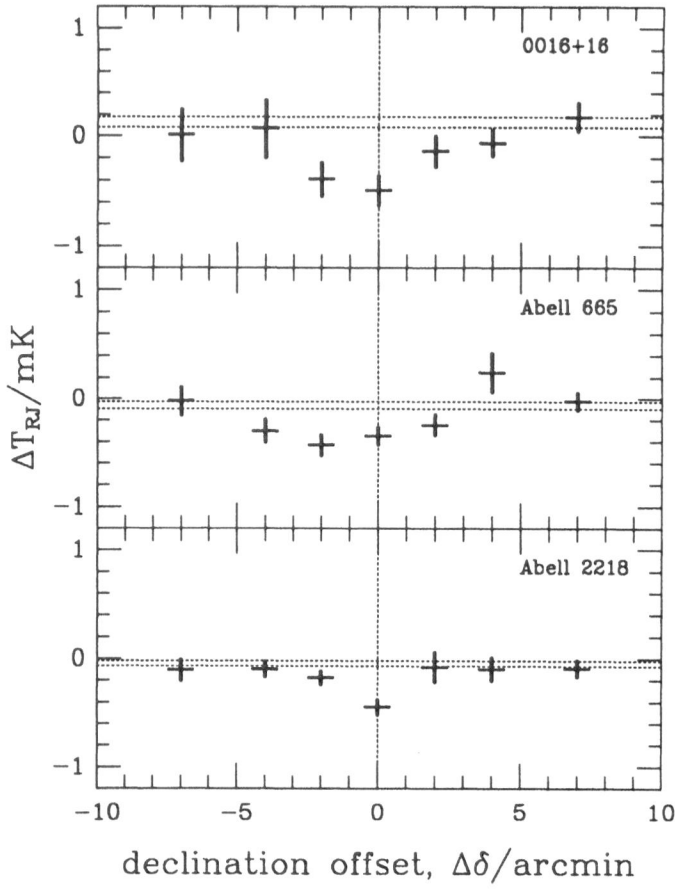

Fig. 3. Measurements of the microwave background radiation as a function of declination near the clusters 0016+16, Abell 665 and Abell 2218. The largest Sunyaev-Zel'dovich effect is seen at the point closest to the X-ray center for each cluster (offset from the scan center in the case of Abell 665), and the apparent angular sizes of the effects are consistent with the predictions of simple models based on the X-ray data (Birkinshaw *et al.* 1993). The horizontal lines delimit the range of possible zero levels, and the errors include both random and systematic components.

multaneous operation in several bands. A suitable choice of differencing between elements of the array reproduces many of the sky-noise subtraction properties of radiometric observing, and the multiband capability holds out the hope of rapid spectral measurements. These same differencing schemes introduce limitations on the selection of clusters that are similar to those that apply to radiometric work, but the smaller angular separations of the beams often causes the minimum redshift cutoff to be rather high, and the peak observing efficiency to be low (as in Chase *et al.* 1987).

The most severe problem with this technique is the extremely high sky brightness against which observations must be made. Coupled with the varying opacity of the sky, this implies that telescopes on high, dry, sites are essential for efficient observing. In the future, space operation with bolometer arrays may provide excellent Sunyaev-Zel'dovich effect data.

This technique is exemplified by the recent work of Wilbanks *et al.* (1993), who used the CSO on Mauna Kea with a three-element array to detect the Sunyaev-Zel'dovich effect from Abell 2163, a cluster of galaxies with an exceptionally hot atmosphere (Arnaud *et al.* 1992) and a bright radio halo source (Herbig & Birkinshaw 1993). The combination of drift-scanning and element-to-element differencing used by Wilbanks *et al.* achieved an excellent separation of the atmospheric signal from the Sunyaev-Zel'dovich effect, and the time-series analysis of their data provides a measurement of the angular structure of the effect. At the wavelength of operation ($\lambda = 2.2$ mm) radio source confusion is not a problem: low-frequency work on Abell 2163 is precluded by the radio environment near the cluster center.

3.3 Interferometric methods

The two techniques discussed above have provided most of the existing data on the Sunyaev-Zel'dovich effect — they are good for surveys of clusters but are suitable only for simple mapping (as in Fig. 3). Interferometry is probably the best method for making detailed images of Sunyaev-Zel'dovich effects.

The major problem with using interferometers to map the microwave background radiation has been that these telescopes are usually designed to provide high sensitivity at high resolution, and hence have large antennas which are widely separated. Observations of clusters of galaxies, on the other hand, demand sensitivity on large angular scales, and structure on these scales is resolved out by large antenna-antenna separations. Thus, for example, the VLA observations of Partridge *et al* (1987) suffered from a factor > 10 suppression of the Sunyaev-Zel'dovich effect signal from Abell 2218 because of the excessive size of the array.

If an interferometer optimized for microwave background work were to be built, it would offer substantial advantages over single-dish radiometers. First, the mix of antenna-antenna separations in an interferometer corresponds to a range of angular scales on the sky: the wider antenna separations are insensitive to the Sunyaev-Zel'dovich effect and can be used as a monitor of confusing radio sources, while the shorter antenna spacings are simultaneously sensitive to the emission from these sources and the Sunyaev-Zel'dovich effect. Second, the instrument would produce a map of the decrement on the sky (on some restricted range of angular scales: although a map of sources and decrement together could theoretically be made). Third, the systematic errors introduced by an interferometer are quite different from those of the other methods, and hence will provide important independent measurements of the effect.

Jones *et al.* (1993) have now used the Ryle interferometer to map the fields of a number of clusters, and have achieved a detection of the Sunyaev-Zel'dovich effect in Abell 2218. This work was done at 15 GHz, and Ryle telescope baselines from 18 to 108 m were used to locate sources and to map the diffuse Sunyaev-Zel'dovich effect. The Sunyaev-Zel'dovich signal seen is roughly consistent with that shown in Fig. 3, but there is a hint of structure between the scales of the 1.8-arcmin beam of the OVRO telescope and the 0.5-arcmin resolution limit of the Ryle data. Even with the small (13-m diameter) antennas and the most compact configuration of the Ryle telescope, the effect is heavily resolved and is detected only on the shortest baselines.

4 Data

Since the first discussion of the Sunyaev-Zel'dovich effect (when it was proposed as a test for the thermal or non-thermal nature of cluster X-ray sources; Sunyaev & Zel'dovich 1972), many searches for the effects have taken place. At present detections at the 4σ confidence level or greater have been reported only for the seven objects listed in Table 2: six clusters of galaxies, and a line of sight towards the quasar PHL 957 that is thought to pass through one or more clusters. The table reports only independent observations of the clusters, eliminating earlier reports based on subsets of the same data used in later papers.

Table 2. Sunyaev-Zel'dovich effect detections at $> 4\sigma$ significance

cluster	redshift	reference
PHL 957 line of sight	$0.4 - 2.5$	Andernach *et al.* 1986
0016+16	0.541	Birkinshaw & Gull 1984
		Birkinshaw *et al.* 1993
Abell 576	0.038	Lake & Partridge 1980
		Birkinshaw *et al.* 1981
Abell 665	0.182	Birkinshaw & Gull 1984
		Birkinshaw *et al.* 1993
Abell 1656 (Coma)	0.023	Herbig *et al.* 1993
Abell 2163	0.170	Wilbanks *et al.* 1993
Abell 2218	0.171	Schallwich 1979
		Birkinshaw *et al.* 1981
		Birkinshaw *et al.* 1993
		Jones *et al.* 1993

The consistency of the measurements has generally been poor. Part of the problem has been the different telescope characteristics that have been used (beam-width, beam-switching angle, etc.), so that a detailed knowledge of the cluster structure is required to compare the results of different observers, but a larger and more serious problem has been the presence of unrecognized systematic errors in the data. Thus for Abell 576, for example, later measurements

(Lasenby & Davies 1983, Birkinshaw & Gull 1984) do not confirm the early reports of large effects (Lake & Partridge 1980, Birkinshaw *et al.* 1981). More recent observations have been in better agreement as a consequence of the increased sophistication forced on observers by the difficulty of the observations.

A rough indication of progress in the measurement of the Sunyaev-Zel'dovich effect can be seen in Table 3, which for Abell 2218 lists the measured results, ΔT_{RJ}, and the implied central decrement at low frequency, ΔT_{RJ0} (calculated using a model for the cluster gas derived from the *Einstein* X-ray image of the cluster; Birkinshaw & Hughes 1993). The table contains only those independent observations that claim errors of better than 0.5 mK in the measured brightness. Abell 2218 is used because it has been observed most frequently. Although many of the recent results in Table 3 are consistent with a value of ΔT_{RJ0} near -0.7 mK, the results are highly discordant. Using the existing Sunyaev-Zel'dovich effect data to draw physical conclusions requires a careful consideration of the severity of the residual systematic errors: it is unwise to use the published measurements uncritically.

Table 3. Independent measurements with error < 0.5 mK towards Abell 2218

reference	reported $\Delta T_{\mathrm{RJ}}/\mathrm{mK}$	inferred $\Delta T_{\mathrm{RJ0}}/\mathrm{mK}$
Perrenod & Lada 1979	$- 1.04 \pm 0.48$	$- 2.83 \pm 1.30$
Schallwich 1979	$- 1.22 \pm 0.25$	$- 2.01 \pm 0.41$
Lake & Partridge 1980	$+ 0.71 \pm 0.38$	$+ 1.89 \pm 1.01$
Birkinshaw *et al.* 1978	$- 1.05 \pm 0.21$	$- 2.90 \pm 0.58$
Birkinshaw & Gull 1984	$- 0.38 \pm 0.19$	$- 1.00 \pm 0.50$
Birkinshaw & Gull 1984	$- 0.31 \pm 0.13$	$- 0.69 \pm 0.29$
Uson 1985	$- 0.29 \pm 0.34$	$- 0.64 \pm 0.53$
Radford *et al.* 1986	$+ 0.16 \pm 0.43$	$+ 0.49 \pm 1.32$
Radford *et al.* 1986	$+ 0.41 \pm 0.32$	$+ 1.16 \pm 0.90$
Klein *et al.* 1991	$- 0.60 \pm 0.20$	$-11.4 \ \pm 3.8$
Birkinshaw *et al.* 1993	$- 0.40 \pm 0.05$	$- 0.62 \pm 0.08$
Jones *et al.* 1993	$- 0.05 \pm 0.01$	$- 1.30 \pm 0.25$

5 Implications

5.1 Hubble constant

If the kinematic effect is small, then the measured Sunyaev-Zel'dovich effect from a cluster of galaxies is proportional to some average of $n_e T_e d$ through a cluster, where the average depends on the structure of the cluster and the properties of the telescope used. Similarly, the measured X-ray flux from the cluster is proportional to some average of $n_e^2 \Lambda(T_e)d$, where Λ is the emissivity of the cluster gas (which depends on the gas temperature and metallicity as well as the energy band observed). X-ray spectral data measure T_e and the metallicity. Thus the two unknown quantities n_e and d, the electron concentration and the path length through the cluster, can be deduced from the observed Sunyaev-Zel'dovich effect and X-ray surface brightness. If the path length is compared with the angular size of the cluster, a measure of the cluster's distance is obtained, and hence the value of the Hubble constant can be measured (Gunn 1979).

This method makes a number of assumptions about the degree to which the structure of the atmosphere can be modeled, and assumes that the cluster atmosphere is spherical (so that the line of sight path length can be compared with the angular size). These are not necessarily good assumptions — in particular, there is a selection effect in favor of clusters elongated in the line of sight (which tend to have the highest surface brightnesses). Clusters of galaxies are nevertheless excellent cosmological probes because they can be seen to large distances, and hence the Hubble constant can be measured on scales ≈ 1 Gpc without recourse to the usual cosmic distance ladder and without the need for significant corrections for local velocity anomalies. Furthermore, each cluster provides an independent measurement of the Hubble constant: there is no need to assume uniformity of clusters, since each can be treated as an individual.

Over the past few years, this method has been applied to two clusters for which excellent X-ray and Sunyaev-Zel'dovich effect data exist, Abell 665 and Abell 2218 (Birkinshaw *et al.* 1991, Birkinshaw & Hughes 1993). Remarkably similar values of the Hubble constant (of about 45 and 50 $\mathrm{km\,s^{-1}\,Mpc^{-1}}$) are obtained, tending to support the 'long' distance scale for the Universe. The error on $H_0 \approx 25$ per cent, dominated by uncertainties in the Sunyaev-Zel'dovich effect data and in the value of T_e.

Although this same method could also be used to measure q_0, the errors are too large for an interesting result to be derived. If q_0 were to be measured to ± 0.5 using the clusters Abell 2218 and 0016+16 then the distance of each would be needed to an accuracy of better than 5 per cent. The 2 per cent error on the Sunyaev-Zel'dovich effect data that this implies is beyond the present observational capabilities.

5.2 CMB structure and spectrum

The distance independence of the Sunyaev-Zel'dovich effect causes the (negative) luminosity from a cluster of galaxies to increase in magnitude with redshift as $L \propto -(1 + z)^4$. If clusters at high redshift were similar to those detected today, then their Sunyaev-Zel'dovich effects would make a significant contribution to radio source confusion and to the small-angular-scale anisotropy in the microwave background radiation, and their integrated effect might cause a significant distortion of the spectrum of the microwave background radiation. The Sunyaev-Zel'dovich effects of superclusters and protoclusters have also been suggested as possible sources of significant brightness fluctuations in the microwave background (Hogan 1992; SubbaRao *et al.* 1993).

The absence of any detectable y parameter in the cosmic microwave background spectrum (Mather *et al.* 1990), and the absence of large numbers of extended negative sources in deep radio surveys (e.g., Fomalont *et al.* 1991), can be used to set limits both to cluster evolution and to the value of q_0 (since q_0 dictates how the volume element in the Universe evolves). This exercise has been conducted by a number of investigators (e.g., Markevitch *et al.* 1992, 1993; Bartlett & Silk 1993). Their results indicate the strong dependence of the predictions on both the value of q_0 and the manner in which cluster atmospheres evolve. The X-ray fading of distant clusters (Edge *et al.* 1990, Gioia *et al.* 1990) provides direct evidence for cluster evolution, and reduces the sensitivity of the method to the volume element evolution at large redshift, but useful limits to the evolution of clusters and q_0 have been set in this way.

5.3 Cluster Properties

The X-ray and Sunyaev-Zel'dovich effect data probe different properties of cluster atmospheres, and so a comparison of these data should provide unique information on the intracluster medium. Attempts to deduce the Hubble constant (Sec. 5.1) have been based on a simple model for cluster atmospheres

$$
\begin{aligned}
n_e &= n_{e0} \left(1 + (r/r_{cx})^2\right)^{-\frac{3}{2}\beta} \\
T_e &= \text{constant}
\end{aligned}
\tag{3}
$$

where variations of the shape parameter β and the scale parameter r_{cx} (or its angular equivalent, θ_{cx}) are sufficient to describe the X-ray surface brightnesses of many clusters. Since the Sunyaev-Zel'dovich data (e.g., Fig. 3) are consistent with the same values of β and θ_{cx} it is of interest to ask what variations from (3) are consistent with the data, but the poor angular resolution of most Sunyaev-Zel'dovich effect data permits only limited statements about structural properties to be made. Note that any structural deviations from (3) (e.g., because of strong clumping of the intracluster medium) will complicate the use of the Sunyaev-Zel'dovich effect as a cosmological probe, but that the agreement between the values of H_0 deduced from two clusters suggests that these structural variations are not large (Birkinshaw & Hughes 1993).

Sunyaev-Zel'dovich effect data can also be used to limit the peculiar velocities of clusters, provided that the value of the Hubble constant is known (e.g., Rephaeli & Lahav 1991). If the smooth isothermal atmosphere model (3) is adopted, then for $H_0 = 50 \text{ km s}^{-1} \text{ Mpc}^{-1}$ it is found that the peculiar velocities of 0016+16, Abell 665, and Abell 2218 are consistent with zero, $|v_r| \lesssim 4000 \text{ km s}^{-1}$. For $H_0 = 100 \text{ km s}^{-1} \text{ Mpc}^{-1}$ these clusters exhibit positive peculiar velocities of several thousand km s^{-1} (Birkinshaw *et al.* 1993). As in Sec. 5.1, these large velocities could be another manifestation of the orientation bias, and spectral data are needed for definitive measurements of v_r. Since microwave background data could also be used to measure the transverse motions of clusters of galaxies (e.g., Birkinshaw 1989), it may be possible in the future to measure the peculiar velocities imposed on the Hubble flow by the formation of large scale structure.

Under the assumption that cluster peculiar velocities are small, and with some choice of H_0, the X-ray and Sunyaev-Zel'dovich effect structural data can be used to estimate the gas temperature, which can be compared with the result obtained by direct X-ray spectroscopy. Differences between these temperatures are possible because the Sunyaev-Zel'dovich effect data are sensitive to the properties of gas in the outer parts of a cluster while the X-ray data are more sensitive to the dense cores of cluster atmospheres, but no evidence for thermal structure has been found. Clusters detected in the Sunyaev-Zel'dovich effect tend to be hotter than average (c.f., Abell 2163; Sec. 3.2), presumably because the strong T_e-dependence of ΔT_{RJ} has led observers to prefer high-T_e clusters.

6 Future prospects

Single dish systems operated at centimeter wavelengths offer an efficient method of searching samples of clusters for the Sunyaev-Zel'dovich effect (especially with the improved sensitivity and stability afforded by modern HEMT-based receivers). Such surveys offer the best method of avoiding selection effects that bias interpretations of the data, and are needed to establish the Sunyaev-Zel'dovich effect as a routine astrophysical tool.

Bolometer arrays continue to improve. The extended spectral grasp of these devices should allow operation beyond the null in the thermal Sunyaev-Zel'dovich effect at 220 GHz, and achieve the spectral separation of the thermal and kinematic effects. The measurement of the radial peculiar velocity of a cluster of galaxies at moderate redshift would allow the study of the evolution of clustering through the changing velocity field. However, space-based systems may be necessary to achieve sufficient sensitivity, and improved radiometers (using modern HEMTs) may displace bolometers as the detectors of choice.

Clusters with suitable angular sizes can now be mapped using radio interferometers, and detailed Sunyaev-Zel'dovich effect images should be available for a number of clusters in the near future. The next step should be the construction of an interferometer customized to the study of the microwave background radiation by having a large number of small antennas arranged in a dense array.

Further substantial progress in observing, detecting, and mapping Sunyaev-Zel'dovich effects requires the development of optimized instruments to replace the general-purpose telescopes presently in use. Improved Sunyaev-Zel'dovich effect data are needed to match the high-quality X-ray data that are now available (Asuka's spectroscopic data, ROSAT's images) or will become available in a few years (AXAF-S's spectroscopy, AXAF-I's images), and which promise to extend our knowledge of cluster atmospheres to substantial redshifts. Such Sunyaev-Zel'dovich effect data will assist in the study of the evolution of clustering and cluster atmospheres, and may map the Hubble flow to redshifts > 1.

References

Andernach, H., Schallwich, D., Sholomitski, G.B., Wielebinski, R.: A search for the microwave diminution towards the cluster 0016+16. Astr. Astrophys. **124** (1983) 326

Andernach, H., Schlickeiser, R., Sholomitski, G.B., Wielebinski, R.: Radio search for the Sunyaev-Zeldovich effect in the vicinity of PHL 957: Astr. Astrophys. **169** (1986) 78

Arnaud, M. *et al.*: A 2163: an exceptionally hot cluster of galaxies. Astrophys. J. **390** (1992) 345

Bartlett, J.G., Silk, J.: The Sunyaev-Zel'dovich effect and cluster evolution. Astrophys. J. (1993) in press

Birkinshaw, M.: Moving gravitational lenses. In Gravitational Lenses (1989) p. 59; eds. Moran, J., Hewitt, J., Lo, K.Y.; Springer-Verlag, Berlin

Birkinshaw, M., Gull, S.F.: Measurements of the gas contents of clusters of galaxies by observations of the background radiation. III. Mon. Not. R. astr. Soc. **206** (1984) 359

Birkinshaw, M., Gull, S.F., Hardebeck, H.: Confirmation of the Sunyaev-Zel'dovich effect towards three clusters of galaxies. Nature, **309** (1984) 34

Birkinshaw, M., Gull, S.F., Hardebeck, H.E., Moffet, A.T.: The structures of the Sunyaev-Zel'dovich effects towards three clusters of galaxies. Astrophys. J. (1993) submitted

Birkinshaw, M., Gull, S.F., Northover, K.J.E.: Measurements of the gas contents of clusters of galaxies by observations of the background radiation at 10.6 GHz. Mon. Not. R. astr. Soc. **185** (1978) 245

Birkinshaw, M., Gull, S.F., Northover, K.J.E.: Measurements of the gas contents of clusters of galaxies by observations of the background radiation at 10.6 GHz. II. Mon. Not. R. astr. Soc. **197** (1981) 571

Birkinshaw, M., Hughes, J.P.: Abell 2218 and the Hubble constant. Astrophys. J. (1993) in press

Birkinshaw, M., Hughes, J.P., Arnaud, K.A.: A measurement of the value of the Hubble constant from the X-ray properties and the Sunyaev-Zel'dovich effect of Abell 665. Astrophys. J. **379** (1991) 466

Chase, S.T., Joseph, R.D., Robertson, N.A., Ade, P.A.R.: A search for the Sunyaev-Zeldovich effect at millimetre wavelengths. Mon. Not. R. astr. Soc. **225** (1987) 171

Edge, A.C., Stewart, G.C., Fabian, A.C., Arnaud, K.A.: An X-ray flux-limited sample of clusters of galaxies: evidence for evolution of the luminosity function. Mon. Not. R. astr. Soc. **245** (1990) 559

Fomalont, E.B., Windhorst, R.A., Kristian, J.A., Kellerman, K.I.: The micro-jansky radio source population at 5 GHz. Astr. J. **102** (1991) 1258

Gioia, I.M. *et al.*: The extended medium sensitivity survey distant cluster sample: X-ray cosmological evolution. Astrophys. J. **356** (1990) L35

Gunn, J.E.: The Friedmann models and optical observations in cosmology. In Observational Cosmology (1978) p. 1; eds Maeder, A., Martinet, L., Tammann, G.; Geneva Obs., Sauverny, Switzerland

Herbig, T., Lawrence, C.R., Readhead, A.C.S., Gulkis, S.: Detection of the Sunyaev-Zel'dovich Effect in the Coma Cluster. Nature (1993) in press

Herbig, T., Birkinshaw, M.: The radio properties of Abell 2163. Astrophys. J. (1993) in preparation

Hogan, C.J.: COBE anisotropy from supercluster gas. Astrophys. J. **398** (1992) L77

Jones, M. *et al.*: Interferometric observation of the Sunyaev-Zel'dovich effect towards Abell 2218. Nature (1993) in press

Klein, U., Rephaeli, Y., Schlickeiser, R., Wielebinski, R.: Measurement of the Sunyaev-Zel'dovich effect towards the A 2218 cluster of galaxies. Astr. Astrophys. **244** (1991) 43

Lake, G., Partridge, R.B.: Microwave search for ionized gas in clusters of galaxies. Astrophys. J. **237** (1980) 378

Lasenby, A.N., Davies, R.D.: λ6-cm observations of fluctuations in the 3 K cosmic microwave background. Mon. Not. R. astr. Soc. **203** (1984) 1137

Markevitch, M. *et al.*: Arcminute fluctuations in the microwave background from clusters of galaxies. Astrophys. J. **395** (1992) 326

Markevitch, M. *et al.*: Cluster evolution and microwave source counts. Astrophys. J. (1993) in press

Mather, J.C, *et al.*: A preliminary measurement of the cosmic microwave background spectrum by the Cosmic Background Explorer (COBE) satellite. Astrophys. J. **354** (1990) L37

McKinnon, M.M., Owen, F.N., Eilek, J.A.: The Sunyaev-Zeldovich effect in radio jet lobes. Astr. J. **101** (1990) 2026

Meyer, S.S., Jeffries, A.D., Weiss, R.: A search for the Sunyaev-Zel'dovich effect at millimeter wavelengths. Astrophys. J. **271** (1983) L1

Moffet, A.T., Birkinshaw, M.: A VLA survey of the three clusters of galaxies 0016+16, Abell 665, and Abell 2218. Astr. J. **98** (1989) 1148

Partridge, R.B., Perley, R.A., Mandolesi, N., Delpino, F.: Preliminary VLA limits on the Sunyaev-Zel'dovich effect in Abell 2218. Astrophys. J. **317** (1987) 112

Perrenod, S.C., Lada, C.J.: The Sunyaev-Zel'dovich effect at 9 millimeters. Astrophys. J. **234** (1979) L173

Radford, S.J.E. *et al.*: A search for the Sunyaev-Zel'dovich effect at $\lambda = 3$ mm. Astrophys. J. **300** (1986) 159

Rephaeli, Y., Lahav, O.: Peculiar cluster velocities from measurements of the kinematic Sunyaev-Zeldovich effect. Astrophys. J. **372** (1991) 21

Schallwich, D.: The Sunyaev-Zel'dovich effect in Abell 2218. Poster presented at IAU Symposium **97** Extragalactic Radio Sources (1979), Albuquerque, NM

SubbaRao, M. *et al.*: Microwave background fluctuations due to the Sunyaev-Zeldovich effect in pancakes. Astrophys. J. (1993) in press

Sunyaev, R.A., Zel'dovich, Ya.B.: The observation of relic radiation as a test of the nature of X-ray radiation from the clusters of galaxies. Comm. Astrophys. Sp. Phys. **4** (1972) 173

Sunyaev, R.A., Zel'dovich, Ya.B.: The velocity of clusters of galaxies relative to the microwave background. The possibility of its measurement. Mon. Not. R. astr. Soc. **190** (1980) 413

Uson, J.M.: The microwave background radiation. Observational and theoretical aspects of relativistic astrophysics and cosmology (1985) 269; ed. J.L. Sanz, L.J. Goicoechea; World Scientific Publishing Co.

Uson, J.M.: The Sunyaev-Zel'dovich effect: measurements and implications. NRAO Greenbank workshop **16** (1987) 255; eds. C. O'Dea, J. Uson; NRAO Greenbank WV

Wilbanks, T.M. *et al.*: Measurement of the Sunyaev-Zel'dovich effect in Abell 2163 at 2.2 mm. Astrophys. J. (1993) in preparation

Clusters and the Cosmic Microwave Background

James G. Bartlett[1] and Joseph Silk[2]

[1] DAEC, Observatoire de Paris-Meudon, 92195 Meudon Cedex, FRANCE
associé au CNRS et à l'Université Paris 7
[2] Astronomy and Physics Departments, and the Center for Particle Astrophysics, University of California, Berkeley, CA 94720, U.S.A.

1 Introduction

The same, hot electrons responsible for the X-ray emission observed from galaxy clusters produces, via inverse Compton scattering, the Sunyaev-Zel'dovich effect, a spectral distortion of the Cosmic Microwave Background (CMB) [1]. The detection of the effect towards individual clusters has proven to be difficult, but there now appear to be several well established examples [2]. With the development of bolometer arrays and the use of dedicated radio telescopes, the Sunyaev-Zel'dovich (SZ) effect should soon become as useful a tool for studying galaxy clusters as X-ray observations have been.

Like the X-ray observations, the SZ effect probes the state of the hot intracluster medium (ICM) and thus the global properties of the cluster gravitational potential. For example, if the gas was heated by infall during cluster formation, then its temperature represents the depth of the potential well. This makes X-ray observations of clusters particularly useful for constraining models of large-scale structure formation. Using a simple approach like Press and Schechter [3], one can write the number density N of clusters per comoving volume as a function of mass as follows:

$$\frac{dN}{dlnM} = \sqrt{\frac{2}{\pi}} \frac{\rho_o}{M} \nu_z(M) \left(- \frac{dln\sigma}{dlnM} \right) e^{-\nu_z^2/2}, \tag{1}$$

where $\rho_o = 1.88 \times 10^{-29}$ g/cm^3 is the current mass density (for the case considered here of a flat universe), $\nu_z(M) = \delta_c(1+z)/\sigma$, $\delta_c = 1.68$, and $\sigma(M)$ is the current-epoch, linearly extrapolated power spectrum smoothed with a top-hat filter on a scale of mass M.

In a flat universe, $T = T_{15}M_{15}^{2/3}(1 + z)$, where the mass of the cluster M_{15} is expressed in terms of 10^{15} solar mass units [4]. The constant T_{15} is given by hydrodynamical simulations to be $6.4h^{2/3}$ keV, with $h \equiv H_o/100$ km/s/Mpc [5]. Thus one can turn equation 1 into the temperature function of galaxy clusters, and by comparing this with data, constrain the amplitude of the density perturbations. This is done in figure 1, where I show the temperature function for two power spectra: $P(k) \propto k^n$ for $n = -1$ and $n = -2$ (all results presented are for $\Omega = 1$ and $h = 1/2$). The data are taken from Henry and Arnaud 1991 [6] (squares) and Edge *et al.* [7] (triangles). The $n = -1$ power spectrum approximates a cold dark matter (CDM) universe on cluster scales, while the $n = -2$ power spectrum resembles, for example, a mixed dark matter (MDM) universe. Notice that the shape of the latter fits that of the temperature function better. Two normalizations are given for the CDM-like spectrum, one corresponding to the COBE amplitude of CMB temperature fluctuations [8], and the other chosen to match the abundance of clusters. Unlike CDM, the $n = -2$ spectrum accounts for *both* the COBE result and the abundance of clusters, making it a currently fashionable model for structure formation [9, 10, 11, 12, 13].

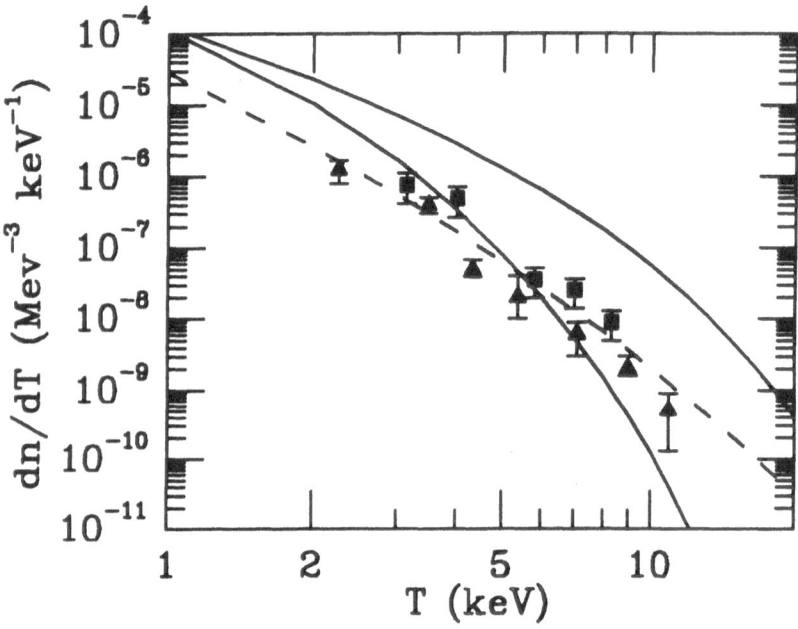

Fig. 1. *Solid lines:* upper-$(n = -1, b = 1)$, lower-$(n = -1, b = 1.7)$ *Dashed line:* $(n = -2, b = 1.7)$ *Data:* see text.

From the above exercise, one sees the utility of the X-ray observations for testing models of galaxy formation. In this contribution, I explore analogous methods employing the SZ effect. Specifically, for the two power spectrum discussed above, I calculate the distribution of Compton y values at different epochs, the number count-flux relation for clusters, and the overall effect of the cluster population on the spectrum and anisotropy of the CMB. With new data on the SZ effect, one hopes that some or all of these quantities will soon prove their worth in constraining models (see, for example, refs. [14, 15, 16, 17, 18, 19, 20]).

2 The ICM Models

In order to proceed and transform equation 1 into observable SZ distributions, such as the Compton y distribution function, we must model the cluster gas density and spatial extent. For simplicity I model these quantities as power laws in the cluster virial mass and redshift:

$$n_g \propto M_{15}^p (1+z)^q \qquad r_c \propto M_{15}^r (1+z)^s. \tag{2}$$

The four exponents here are not completely independent. The X-ray luminosity may be written as $L \sim n_g^2 T^{1/2} r_c^3 \sim M_{15}^P (1+z)^Q$, where $P = 2p + 3r + 1/3$ and $Q = 2q + 3s + 1/2$. One expects, if the gas has not cooled or been significantly heated after infall, that in a flat universe $(p, q) = (0, 3)$ and $(r, s) = (1/3, -1)$, and hence $P = 4/3$ and $Q = 7/2$, the so-called "self-similar" case [4]. One can attempt to determine P and Q from the observations. For example, the observed correlation $L \sim T^{2.75}$ for the local cluster population implies that $P = 11/6$. Note that this rules out the simple self-similar value. Constraining Q in a similar manner requires temperature measurements at larger redshifts. As these do not yet exist, I will consistently use $Q = 7/2$ in the following.

To explore the importance of the somewhat unknown ICM physics on the SZ results, I consider three models for each power spectrum. Model A will be a purely self-similar model, ignoring the observed $L - T$ relation. The other two models, however, will adopt $P = 11/6$, satisfying the correlation, and $Q = 7/2$. Of these, Model B will also employ $(r, s) = (1/2, -1)$, which allows the gas density to scale with the mean background density, while Model C will use $(r, s) = (1/3, -1)$. The models are then completely specified.

3 The Compton y Distribution Function

The Compton y parameter is an integral of the ICM pressure along the line of sight: $y \sim n_g T r_c$. Equations 2 then allow us to transform the cluster mass function into a distribution of Compton y values for each model. To normalize the relation between y and the the mass and redshift, I use the observed properties of A665, one of clusters with a SZ detection [21]. At a redshift of 0.18, this cluster has a $T \sim 8$ keV, a $r_c = 0.2h^{-1}$ Mpc, and a measured $y = 1.68 \times 10^{-4}$. The result for each model is shown in figure 2 for $z = 0$ and $z = 1$. Notice that for $z = 0$ the normalization and shape of this function depend primarily on the spectral index n, offering a new way to constrain the power spectrum. In addition, we observe that the CDM-like scenarios all display *positive* evolution towards the higher redshifts, exactly opposite to the case of the $n = -2$ spectrum. Perhaps this too can be used to constrain the power spectrum, especially since it is no more difficult to obtain the y distribution at large z than it is at $z = 0$, in contrast to a measurement of the X-ray temperature function.

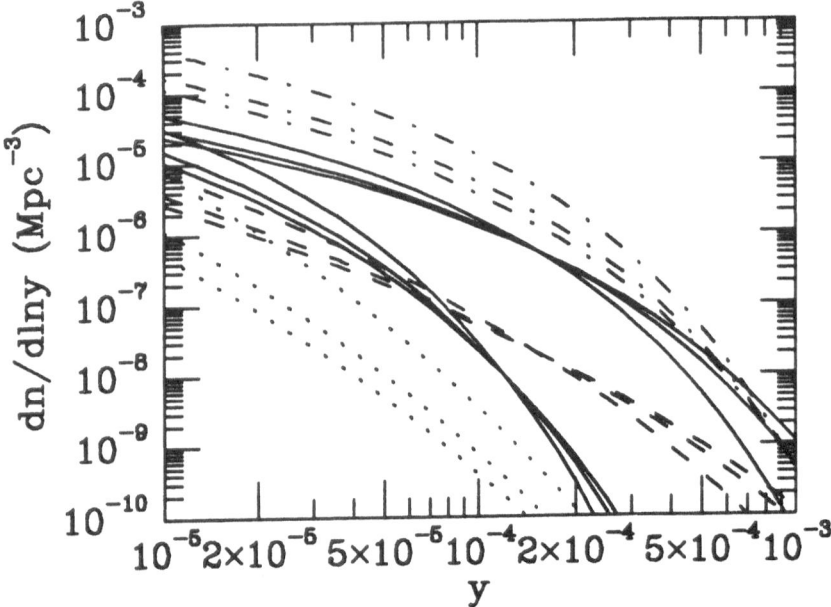

Fig. 2. *Solid lines*: upper - $(z = 0, n = -1, b = 1)$, lower - $(z = 0, n = -1, b = 1.7)$ *Dot-dashed lines*: $(z = 1, n = -1, b = 1)$ *Dashed lines*: $(z = 0, n = -2, b = 1.7)$ *Dotted lines*: $(z = 1, n = -2, b = 1.7)$

4 Counts

Depending on the observation frequency, clusters will appear as either sources of radio emission or as decrements relative to the mean CMB brightness. In either case, one can calculate the number counts as a function of signal strength. To do so, I adopt the "isothermal-β" model for the spatial distribution of the ICM and use r_c as the core radius. The Kompaneets equation [22] gives the surface brightness profile which can be integrated to obtain the total flux density. I impose a cutoff at $5r_c$ for the results here. After normalizing all the relations to A665 and transforming the mass function into a flux density distribution function, I integrate over redshift to obtain the counts shown in figure 3. The turn-down at small flux density is caused by the loss of gas to cooling in the smaller objects. Observing such counts requires large sky coverage, but may never-the-less provide useful constraints on cluster evolution models.

5 Effects on the CMB

Here we study the effects of the entire cluster population on the spectrum and anisotropy of the CMB. The combined effect of all clusters is to produce a mean y distortion in the CMB spectrum. This can be calculated easily by integrating the source counts discussed above. The largest distortion occurs with the unbiased

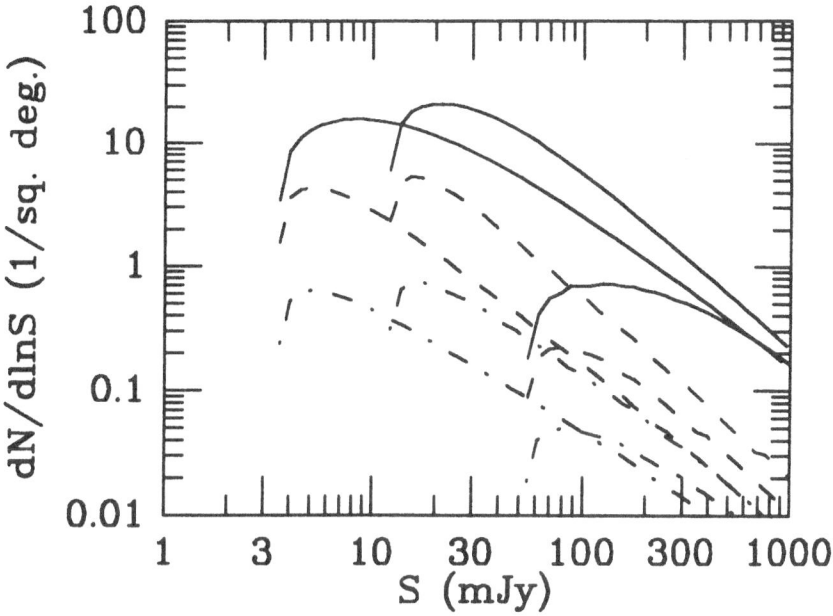

Fig. 3. *Solid lines*: $(n = -1, b = 1)$, *Dashed lines*: $(n = -1, b = 1.7)$ *Dot-dashed lines*: $(n = -2, b = 1.7)$ In all cases, Model B turns down at the smallest S, followed by Model A and finally by Model C.

CDM-like power spectrum employing Model A, for which $\bar{y} = 6 \times 10^{-6}$, a value lower than COBE's current limit of $y < 2.5 \times 10^{-5}$ [23]. The other models fall anywhere from a factor of 10 to 100 below the COBE limit. Thus it seems unlikely that one can use measurements of the CMB spectrum to constrain models of cluster evolution, at least if $\Omega = 1$.

Additionally, the clusters produce an anisotropy in the CMB, and here there appears to be more potential for observing the effect. In figure 4, I show the *rms* temperature fluctuations generated by the clusters for a single gaussian beam as a function of the beam FWHM in arcminutes. The cluster centers are assumed to be uncorrelated for this calculation. These numbers are meant to be representative of the amplitude expected. One cannot interpret them as the usual standard deviation of gaussian fluctuations because the cluster induced perturbations are *not* gaussian in general, being dominated by rare, bright events. This is clear, for example, from the slope of the number counts calculated above. In any case, it does appear that unbiased CDM-like models produce rather excessively large anisotropies given the OVRO limit of $\delta T/T < 1.7 \times 10^{-5}$ for gaussian fluctuations at 20 GHz [24]. There is a slightly more stringent limit from the AT [25].

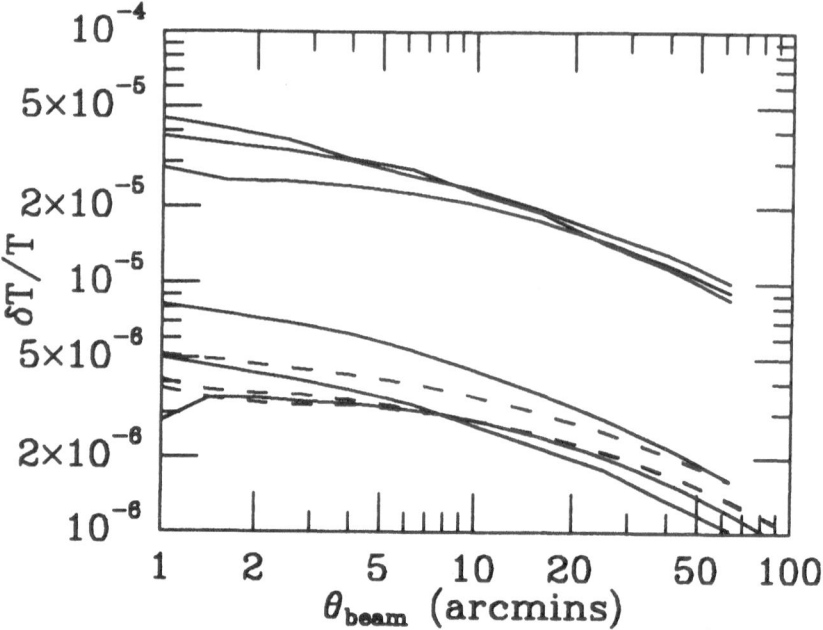

Fig. 4. *Solid lines*: upper - $(n = -1, b = 1)$, lower - $(n = -1, b = 1.7)$ *Dashed lines*: $(n = -2, b = 1.7)$ The numbers represent *rms* values.

6 Conclusion

With new detectors and dedicated efforts, the SZ effect should soon become a useful tool for probing cluster evolution and structure formation. In particular, the y distribution function may be useful for constraining the power spectrum of density perturbations, both by observing the distribution at the current epoch and by determining the sense of its evolution with redshift. Observations of the CMB anisotropy at arcminute scales also appear to be a powerful way to constrain models of structure formation. We found that, for example, unbiased CDM-like scenarios result in rather large anisotropies, perhaps in violation of existing limits. This later statement must be explored further by adequately modeling the nongaussian nature of the cluster induced perturbations.

Although we only considered the case of a flat universe here, one must not overlook the value of the SZ effect for studying an open universe. Because of its unique distance independence, one can observe distant clusters with the same ease as those nearby. This is a significant advantage over the X-ray observations, which suffer from a lack of photons from distant clusters. Thus the SZ effect can be used to search for the high redshift clusters expected in an open universe.

References

1. Sunyaev, R.A., and Zel'dovich, Ya. B. 1972, *Comm. Astropys. Sp. Phys.*, **4**, 173.
2. Birkinshaw, M. 1990, in *The Cosmic Microwave Background: 25 Years Later*, eds. N. Mandolesi and N. Vittorio (Kluwer:Dordrecht).
3. Press, W.H., and Schechter, P. 1974, *Ap. J.*, **187**, 425.
4. Kaiser, N. 1986, *M.N.R.A.S.*, **219**, 785.
5. Evrard, A.E. 1990, in *Clusters of Galaxies*, eds. M. Fitchett and W. Oegerle (Cambridge; Cambridge University Press).
6. Henry, J.P., and Arnaud K.A. 1991, *Ap. J.*, **372**, 410.
7. Edge, A.C., Stewart, G.C., Fabian, A.C., and Arnaud, K.A. 1990, *M.N.R.A.S.*, **245**, 559.
8. Smoot, G.F. *et al.* 1992, *Ap. J. Lett.*, **396**, L1.
9. Bartlett, J.G., and Silk, J. 1993, *Ap. J. Lett.*, **407**, L45.
10. Schaefer, R.K., and Shafi, Q. 1992, *Nature*, **359** 199.
11. Davis, M., Summers, F.J., Schlegel, D. 1992, *Nature*, **359**, 393.
12. Taylor, A.N., and Rowan-Robinson, M. 1992, *Nature*, **359**, 396.
13. Klypin, A., Holtzman, J., Primack, J., Regos, E. 1993, preprint.
14. Korolev, V.A., Sunyaev, R.A., and Yakubtsev., L.A. 1986, *Sov. Astron. Lett.*, **12**, 141.
15. Cole, S., and Kaiser, N. 1989, *M.N.R.A.S.*, **233**, 637.
16. Schaeffer, R., and Silk, J. 1988, *Ap. J.*, **333**, 509.
17. Bond, J.R., and Myers, S.T. 1991, in *Trends in Astroparticle Physics*, ed. D. Cline (Singapore: World Scientific).
18. Cavaliere, A., Menci, N., and Setti, G. 1991, *Astron. Astrophys.*, **245**, L21.
19. Markevitch, M., Blumenthal, G.R., Forman, W., Jones, C., and Sunyaev, R.A. 1991, *Ap. J. Lett.*, **378**, L33.
20. Markevitch, M., Blumenthal, G.R., Forman, W., Jones, C., and Sunyaev, R.A. 1992, *Ap. J.*, **395**, 326.
21. Birkinshaw, M., Hughs, J.P., and Arnaud, K.A. 1991, *Ap. J.*, **379**, 466.
22. Kompaneets, A. 1957, *Sov. Phys.-JETP*, **4**, 730.
23. Mather, J. *et al.* 1993, preprint.
24. Readhead, A.S.C., Lawrence, C.R., Myers, S.T., Sargent, W.L.W., Hardebeck, H.E., and Moffet, A.T. 1989, *Ap. J.*, **346**, 566.
25. Subrahmanyan, R, Ekers, R.D., Sinclair, M., and Silk, J. 1993 *M.N.R.A.S.*, in press.

Theoretical aspects of the CMB spectrum

Luigi Danese[1] and Carlo Burigana[2]

[1] Dipartimento di Astronomia, Vicolo dell'Osservatorio 5, I-35122 Padova, Italy
[2] Osservatorio Astronomico, Vicolo dell'Osservatorio 5, I-35122 Padova, Italy

1 Introduction

If the simple hot Big Bang model is assumed, the CMB is expected to emerge from the early universe exhibiting a black–body (BB) spectrum. Indeed before the end of the lepton era the high number density of electrons and positrons ensures full thermal equilibrium between matter and photons in the Universe. After the annihilation of electron pairs CMB photons interact with a plasma composed by protons, He nuclei (about 23% in weight) and electrons, whose number density n_e reduces to $\sim 10^{-10} \div 10^{-9}$ times the photon number density n_γ. This fact together with the decrease of the particle density with the expansion makes harder and harder the thermodynamical equilibrium between photons and plasma at increasing times.

A nice consequence is that energy injections in the radiation field after a fiducial epoch z_{therm} leave tracks on CMB spectrum. Indeed in pioneering works Weymann (1966), Zeldovich and Sunyaev (1969), Sunyaev and Zeldovich (1970a), Peebles (1971) pointed out that CMB spectrum is a unique probe of physical processes that possibly have occurred in the early universe.

2 Physical processes in the primeval plasma

After the electron-pair annihilation and before the recombination epoch the equilibrium established between baryonic matter and radiation mainly depends on three processes: the Compton scattering (C), the bremsstrahlung (B) and the radiative Compton (RC). While the former process conserves the photon number, the latter processes produce and absorb photons. Their joined action change the photon occupation number $\eta(\nu, t)$ with the time (Danese and De Zotti 1977):

$$\frac{\partial \eta}{\partial t} = \frac{\partial \eta}{\partial t}\mid_C + \frac{\partial \eta}{\partial t}\mid_B + \frac{\partial \eta}{\partial t}\mid_{RC} . \tag{1}$$

Kompaneets (1956) solved the general problem for the Compton scattering passing from the collisional integral equation to a very simple kinetic equation, which reads

$$\frac{\partial \eta}{\partial t}|_C = \frac{1}{t_C} \frac{1}{x_e^2} \frac{\partial}{\partial x_e} \left[x_e^4 \left(\frac{\partial \eta}{\partial x_e} + \eta + \eta^2 \right) \right],$$ (2)

where $x_e = h\nu/(kT_e)$ is a dimensionless frequency; the characteristic time t_C is

$$t_C = t_{\gamma e} \frac{mc^2}{kT_e} \simeq 4.5 \times 10^{28} \left(T_o/2.7\,K \right)^{-1} \phi^{-1} \widehat{\Omega}_b^{-1} (1+z)^{-4} \text{ s},$$ (3)

where $t_{\gamma e} = 1/(n_e \sigma_T c)$ is the photon–electron collision time, T_e is the electron temperature, $T_r = T_o(1 + z)$ is the radiation temperature ($\epsilon_{ro} = aT_o^4$ is the present radiation energy density) and $\phi = T_e/T_r$; kT_e/mc^2 is the mean fractional change of photon energy in a scattering of cool photons off hot electrons ($T_e \gg T_r$); Ω_b is the baryon density in units of the critical density, H_o is the Hubble constant and $\widehat{\Omega}_b = (H_o/50)^2 \Omega_b$.

The bremsstrahlung term is given by

$$\frac{\partial \eta}{\partial t}|_B = \frac{1}{t_B} g_B(x_e) \frac{\exp(-x_e)}{x_e^3} \left[1 - \eta \left(\exp(x_e) - 1 \right) \right];$$ (4)

a quite similar expression holds for the radiative Compton term

$$\frac{\partial \eta}{\partial t}|_{RC} = \frac{1}{t_{RC}} g_{RC}(x_e) \frac{\exp(-x_e)}{x_e^3} \left[1 - \eta \left(\exp(x_e) - 1 \right) \right];$$ (5)

$g_B(x_e)$ and $g_{RC}(x_e)$ are the Gaunt factor of bremsstrahlung and of radiative Compton respectively.
In terms of cosmological parameters the rates at which the bremsstrahlung and radiative Compton processes are able to change the occupation number are given by

$$K_B = t_B^{-1} \simeq 2.6 \times 10^{-25} \left(T_e/T_r \right)^{-7/2} \left(T_o/2.7\,K \right)^{-7/2} (1+z)^{5/2} \widehat{\Omega}_b^2 \text{ s}^{-1}$$ (6)

and

$$K_{RC} = t_{RC}^{-1} \simeq 0.82 \times 10^{-39} \left(T_o/2.7\,K \right)^2 \left(T_e/T_r \right)^2 \frac{\epsilon_r}{aT_e^4} \widehat{\Omega}_b (1+z)^5 \text{ s}^{-1}.$$ (7)

The importance of the radiative Compton respect to the bremsstrahlung is increasing with increasing redshift and decreasing $\widehat{\Omega}_b$, because the radiative Compton depends on n_e and on the radiation energy density whereas the bremsstrahlung depends on n_e^2. For instance if $\widehat{\Omega}_b \simeq 0.1$ the radiative Compton is more efficient than bremsstrahlung at $z \gtrsim 2.5 \times 10^5$.

2.1 The thermalization epoch

In absence of strong density fluctuations in the baryonic component and of copious production of photons (e.g. via particle decay), the three processes mentioned above are able to keep baryonic matter and radiation in thermal equilibrium and produce a BB spectrum even in presence of large energy injections only down to a redshift $z \sim 10^6 \div 10^7$.

After an energy injection in the radiation field a BB spectrum can be obtained only if the photon number density can be adjusted to the new energy density, because energy and photon number density depend only on the BB temperature. Thus the processes of photon production and absorption dictate the time to reach the equilibrium and the timescale is given by

$$t_{therm} \sim n_{BB} \left[\frac{\partial n}{\partial t} \Big|_S + \frac{\partial n}{\partial t} \Big|_B + \frac{\partial n}{\partial t} \Big|_{RC} \right]^{-1} , \qquad (8)$$

where n_{BB} is the photons number density of the black–body at temperature T_e (De Zotti 1986). The term $(\partial n/\partial t)|_S$ accounts for possible photon sources different from bremsstrahlung and radiative Compton.

If $\hat{\Omega}_b \lesssim 0.3$ the thermalization redshift for small distortions is given by

$$z_{therm} \simeq 2.54 \, 10^6 \left(\frac{\Delta \epsilon}{\epsilon_i} \right)^{0.11} \hat{\Omega}_b^{-0.39} \left(\frac{T_0}{2.7 \, K} \right)^{0.25} , \qquad (9)$$

while for large distortions

$$z_{therm} \simeq 2.90 \, 10^6 \hat{\Omega}_b^{-0.36} , \qquad (10)$$

where $\Delta \epsilon/\epsilon_i$ is the fractional amount of energy injected in the radiation field (ϵ_i being the radiation energy density before the heating). The above formulae are a rather accurate description of numerical calculations, in the hypothesis of energy injection with negligible additional photons (Burigana et al. 1991a).

Estimates of z_{therm} have been recently worked out by Hu and Silk (1993) and by Burigana et al. (1993) even in the case of non negligible amounts of extra photons. In the case that the extra photons are significantly less than those required to produce a BB spectrum with the increased energy density, the epoch of thermalization z_{therm} practically does not change. If with the additional photons the photon number density exceeds that required to produce a BB spectrum, then a small decrease of z_{therm} has been found. In the very special case in which energy and photons are added just in the amounts required to produce a BB spectrum, Burigana et al. (1993) have shown that distortions of the CMB are negligible down to redshifts lower by almost a factor of 20 than those given by eqq. (9) and (10).

After z_{therm} the efficiencies of bremsstrahlung and radiative Compton decline and the full equilibrium, if perturbed, can be no longer re-established. However kinetic equilibrium between matter and radiation is achieved through Compton scattering.

3 Evolution of Bose–Einstein spectra

The effectiveness of Compton scattering in maintaining kinetic equilibrium is well represented by the usual dimensionless comptonization parameter (used in the following as a dimensionless time parameter)

$$y_e = \int_0^z \frac{dz}{z} \frac{t_{exp}}{t_C} . \tag{11}$$

As it is well known, kinetic equilibrium yields a Bose–Einstein (BE) spectrum, if photon production is neglected. The photon occupation number η is then a function of the temperature and of the chemical potential μ and reads

$$\eta = \eta_{BE} = [\exp(x_e + \mu) - 1]^{-1} . \tag{12}$$

In particular the radiation energy density is

$$\epsilon = aT_i^4 \left(1 + \frac{\Delta\epsilon}{\epsilon_i}\right) = aT_e^4 f(\mu), \tag{13}$$

with T_i the radiation temperature before the injection of energy. The photon number density can be written as

$$n_{BE} = \frac{aT_e^3}{2.7k} \phi(\mu) = \frac{aT_i^3}{2.7k} . \tag{14}$$

Simple expressions exist for both $f(\mu)$ and $\phi(\mu)$:

$$f(\mu) =\simeq \begin{cases} 1 - 1.11\mu & \text{if } \mu \ll 1 \\ 0.924\exp(-\mu) & \text{if } \mu \gg 1 \end{cases}, \tag{15}$$

$$\phi(\mu) =\simeq \begin{cases} 1 - 1.38\mu & \text{if } \mu \ll 1 \\ 0.832\exp(-\mu) & \text{if } \mu \gg 1 \end{cases}. \tag{16}$$

The chemical potential μ is a measure of the fractional amount of extra energy injected in the radiation field. In particular for small distortions, $\mu \ll 1$, $\Delta\epsilon/\epsilon_i \simeq 0.7\mu$.

Kinetic equilibrium is well maintained as long as $t_C \ll t_{exp}$, or, more precisely, $y_e(z) > y_e(z_1) \equiv y_1 \approx 4$ (Zeldovich and Sunyaev 1969; Illarionov and Sunyaev 1974; Chan and Jones 1975; Burigana et al. 1991a). Assuming $\phi = T_e/T_r =$const, we get

$$z_1 \simeq 4.3 \times 10^4 \left(\frac{T_0}{2.7\,K}\right)^{1/2} \left(\frac{\kappa'}{1.68}\right)^{1/4} \phi^{-1/2} \left(\frac{y_1}{\widehat{\Omega}_b}\right)^{1/2} , \tag{17}$$

where $\kappa' = 1 + (\kappa - 1)/(1 + \Delta\epsilon/\epsilon)$ includes the effect of relativistic neutrinos on the expansion time in presence of energy injected in the radiation field (for 3 species of massless neutrinos, $\kappa = 1.68$).

On the other hand the photon emitting and absorbing processes can not be neglected, particularly at $z \geq z_1$ and at low frequencies $x_e \ll 1$; as a consequence the occupation number $\eta(x, t)$ can in first approximation be described by a BE

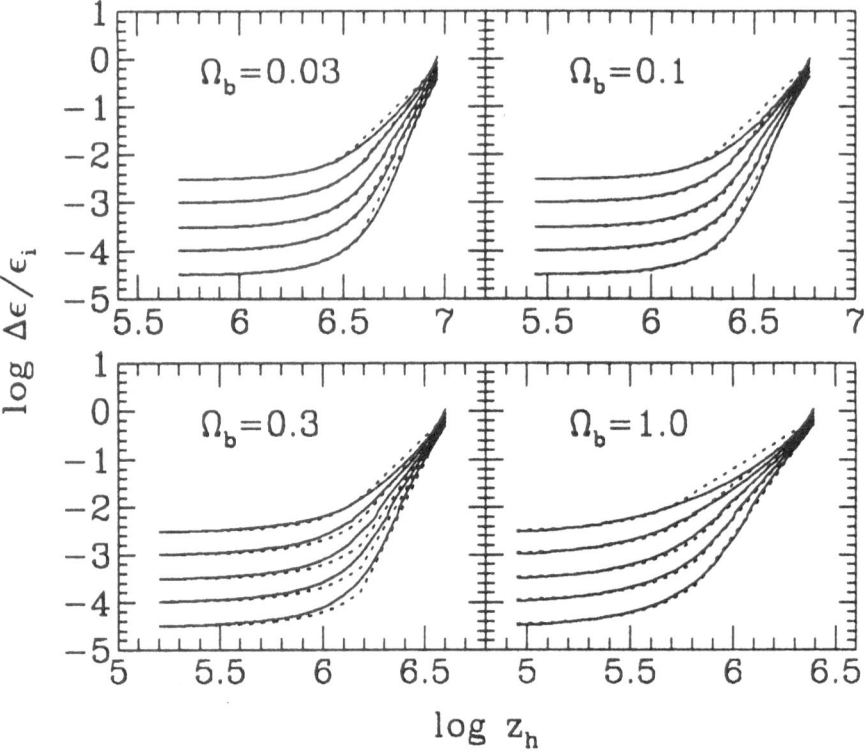

Fig. 1. Fractional amount of energy dissipated at redshift z_h ($z_1 < z_h < 0.9z_{therm}$), corresponding to several present values of the chemical potential, μ_0 [$\mu_0 \simeq 1.4(\Delta\epsilon/\epsilon_i)_{z_1}$], for mean baryon densities showed in each panel ($H_0 = 50$ km s^{-1} Mpc^{-1}). The solid lines show the numerical results; the dashed lines, an useful analytical approximation (Burigana et al. 1991b). Spectral distortions are substantially damped between z_{therm} and z_1, so that even significant amounts of energy released at early times produce only small observable effects at present time.

formula, but with a frequency-dependent chemical potential (Sunyaev and Zeldovich 1970a)

$$\mu(x_e) = \mu_0 \exp\left[-x_c(z)/x_e\right] . \tag{18}$$

The characteristic dimensionless frequency x_c is a function of time and is given by

$$x_c(z) = [K_B g_B(x_c) + K_{RC} g_{RC}(x_c)]^{1/2} t_C^{1/2} . \tag{19}$$

The critical frequency x_c depends on the mean baryon density $\widehat{\Omega}_b$ through K_B, K_{RC} and t_C.

Thus a substantial damping of spectral distortions occurs between z_{therm} and z_1 (see Fig. 1). Moreover bremsstrahlung and/or radiative Compton produce and absorb photons down to z_{rec}, significantly modifying BE like spectra possibly formed before z_1. An analytical approximation that takes into account all the

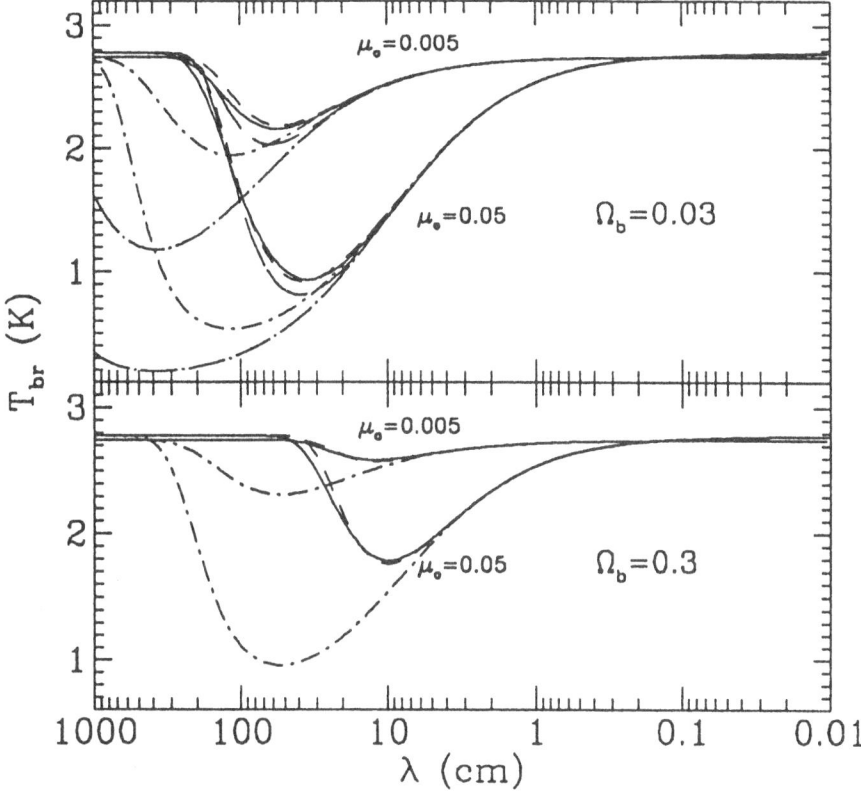

Fig. 2. Comparison between the results of numerical integration of the Kompaneets equation (solid lines), the analytic approximation of Burigana et al. (1991a) (dashed lines) and the original approximation derived by Sunyaev and Zeldovich (1970a) neglecting photon production processes for $z < z_1$ (dot–dashed curves); two values of Ω_b ($H_0 = 50$, $\Omega_T = 1$) and of the chemical potential are considered. Lines characterized by long dashes and dots plus long dashes (top panel only) refer respectively to these two analytic approximation when radiative Compton is neglected.

effects has been derived by Danese and De Zotti (1980) and improved by Burigana et al. (1991a). The significant improvement respect to previous approximations is mainly due to a more detailed analysis of the competition between photon emission processes and Compton scattering. Indeed the maximum frequency at which bremsstrahlung and radiative Compton are able to establish a BB spectrum before the photons are removed by Compton scattering is attained at a redshift $z_p < z_1$

$$z_p \simeq 2.14 \times 10^4 \left(\frac{T_0}{2.7\,K} \right)^{1/2} \left(\frac{\kappa'}{1.68} \right)^{1/4} \widehat{\Omega}_b^{-1/2} . \tag{20}$$

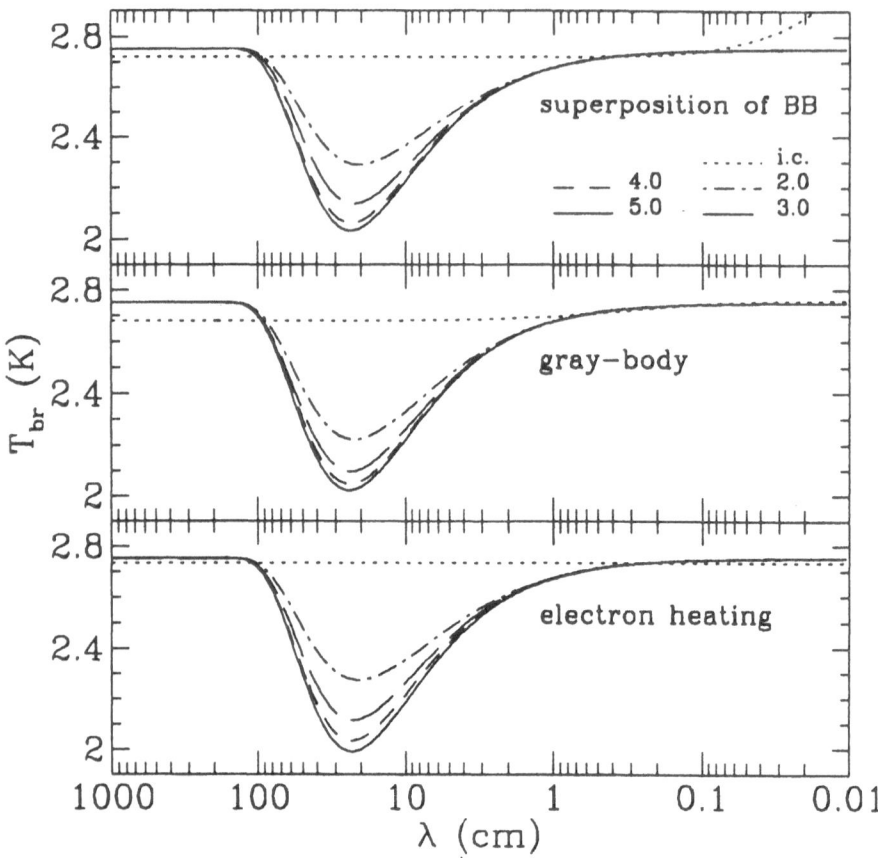

Fig. 3. Relaxation to a Bose–Einstein like spectrum of early distortions corresponding to an energy release $\Delta\epsilon/\epsilon_i = 0.01$ for three kinds of distorted spectra and for several choices of the redshift of heating, z_h, i.e. for the corrisponding value of y_e indicated in the top panel. In the two first cases the energy is injected directly in the radiation field and a superposition of black-body spectra or a gray body spectrum is assumed as initial spectrum (labelled i.c.). In the last case the initial spectrum is a black-body and the energy is released by an external source that heats the matter. The time required to achieve the kinetic equilibrium and the final spectrum for early distortions (i.e. for $y_e(z_h) \gtrsim 2$) are essentially the same for the three processes ($\Omega_b = 0.1$, $H_0 = 50$, $\Omega_T = 1$).

This fact is properly included in the approximation proposed by Burigana et al. (1991a), which quite well fits the numerical solutions (see Fig. 2).

Detailed numerical calculations have been done for a number of possible processes trasferring energy to the radiation field.

The cases presented in Fig. 3 clarify that for $y_e \gtrsim 2$ the final spectrum depends only on the time and amount of energy injection and not on the detailed mechanism of energy injection.

As it is apparent from Figg. 2 and 3, the amplitude of the minimum of the brightness temperature $\Delta T/T$ and its location in wavelength significantly depend on the baryon density and not on the particular process, as first pointed out by Sunyaev and Zeldovich (1970a). Good analytical approximations to detailed numerical calculations give

$$\lambda_m \simeq 5.64 \widehat{\Omega}_b^{-2/3} \text{ cm} \tag{21}$$

and

$$\left(\frac{\Delta T}{T}\right)_m \simeq 5.82\mu_0 \widehat{\Omega}_b^{-2/3} . \tag{22}$$

Thus detection of spectral distortions generated at epochs $z_1 \leq z \leq z_{therm}$ would be much valuable for estimating the mean baryon density.

Unfortunately the hope of detecting BE distortions has been significantly reduced by the results of COBE, because FIRAS data (Mather et al. 1993) suggest that the chemical potential at z_1 is very small:

$$\mu_0 = (-1.2 \pm 1.1) \times 10^{-4} \tag{23}$$

(1σ errors). On the other hand it is well known that the best–fit temperature of the long–wavelength measurements is about 80 mK below the value found by FIRAS. This fact calls for further investigations on the long–wavelength portion of the CMB spectrum.

Because the limits on BE like distortions are narrowed down in a such dramatic way, in the following we shall explore to what extent they constrain some of the energy injection processes occurring at epochs $z_1 < z_h < z_{therm}$.

3.1 Radiative decay of massive particles

We assume that particle decay processes $X \rightarrow X' + \gamma$ create at the decay redshift z_D photons with dimensionless energy

$$x_X = h\nu_D/kT_r(z_D) \sim \frac{1}{2}m_X c^2/kT_r(z_D) . \tag{24}$$

The resulting total energy density is

$$\epsilon_D \simeq \frac{1}{2}m_X c^2 B_\gamma n_X , \tag{25}$$

where B_γ is the branching ratio in the radiative channel and $n_X = (3/8)(g_f/\chi)n_i = R_X n_i$; n_i is the photon number density in the radiation field before the decay, g_f is the number of states per momentum mode and χ is the effective number of relativistic interacting species at the decay epoch (the above formulas are properly correct for non relativistic particles but they are instructive also in the case of relativistic ones). As a consequence we find

$$\frac{\Delta\epsilon}{\epsilon_i} = \frac{x_X R_X B_\gamma}{\overline{x}_{CMB}} \tag{26}$$

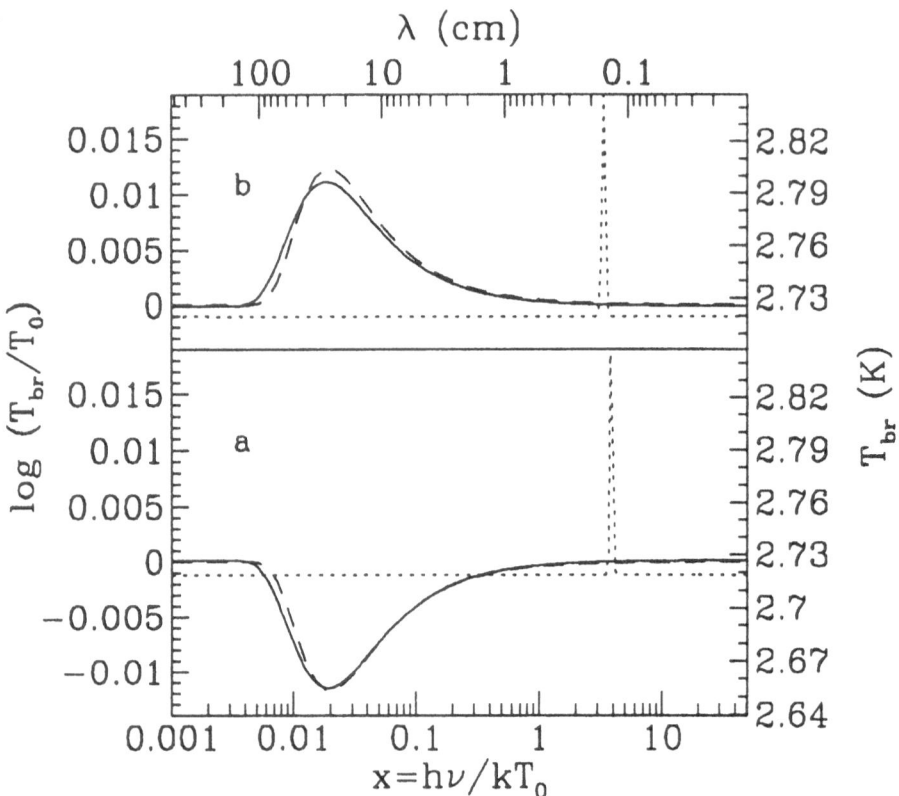

Fig. 4. Relaxation to a Bose–Einstein like spectrum of early distortions in presence of radiative decay with $\Delta n_\gamma/n_i = 7.5\ 10^{-3}$ and $\Delta\epsilon/\epsilon_i$ such that $\mu = 10^{-3}$ (a) and $\mu = -10^{-3}$ (b) (see eq. (28), $\Delta\epsilon/\epsilon_i \simeq 0.01$). The initial spectrum is a black-body plus a "line" due to the radiative decay (dotted lines). The numerical results for the present spectrum (solid lines) and the approximation of Burigana et al. (1991a) (dashed lines) are showed. The agreement results to be quite good ($\Omega_b = 0.1$, $H_0 = 50$, $\Omega_T = 1$).

and

$$\frac{\Delta n_\gamma}{n_i} = R_X B_\gamma , \qquad (27)$$

where $\bar{x}_{CMB} = 2.7$.

Using eqq. (13) and (14) for BE like spectra in the case $\mu \ll 1$ we obtain (Burigana 1989; Hu and Silk 1993)

$$\mu \simeq 1.4 \left(\frac{1 + \Delta\epsilon/\epsilon_i}{(1 + \Delta n_\gamma/n_i)^{\frac{4}{3}}} - 1 \right) = 1.4 \left(\frac{1 + R_X B_\gamma x_X/\bar{x}_{CMB}}{(1 + R_X B_\gamma)^{\frac{4}{3}}} - 1 \right) . \qquad (28)$$

Equations (26) ÷ (28) show that positive and negative values of μ are possible, depending on the parameters of the decay process. Recently Hu and Silk (1993)

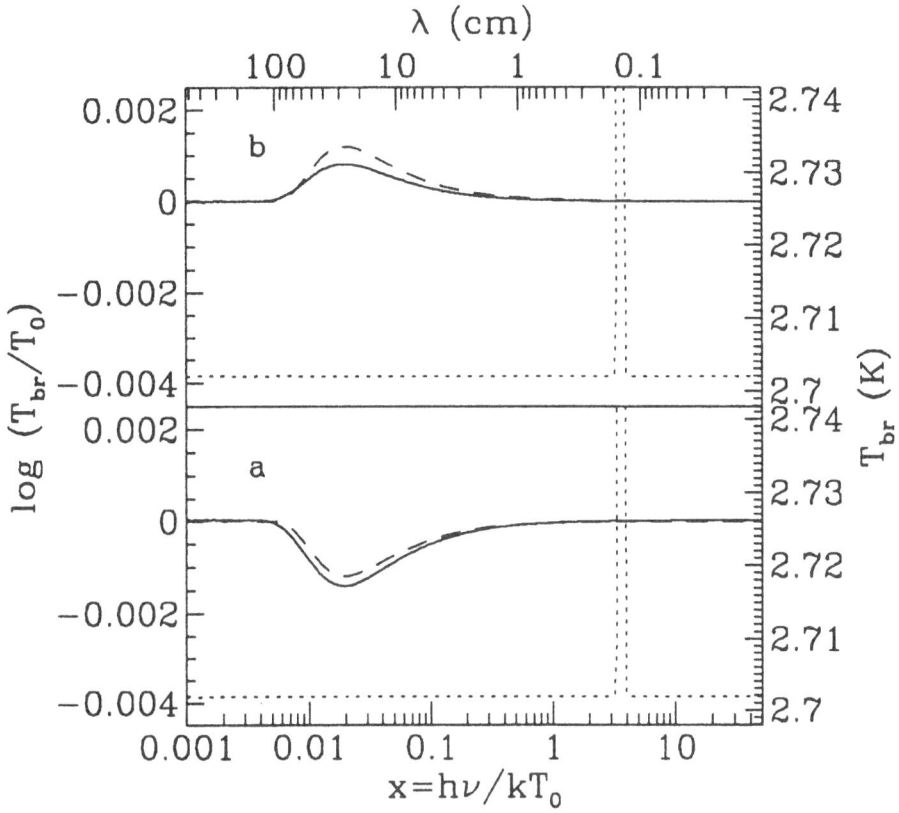

Fig. 5. Same as in Fig. 4, but for a radiative decay characterized by $\Delta n_\gamma/n_i = 0.027$ and $\mu = 10^{-4}$ (a) and $\mu = -10^{-4}$ (b). In these cases the agreement between the numerical results and the analytic approximation is less good than in the cases of Fig. 4, due to the larger number of photons injected in the radiation field.

have explored cases with $\mu < 0$, already partly examined by Kawasaki and Sato (1987).

It is also apparent that μ could vanish even in presence of energy injection $\Delta\epsilon/\epsilon_i > 0$, although the case requires peculiar combinations of m_X, n_X and B_γ. Figg. 4a and 5a show that the cases with $\mu > 0$ are well described by the usual analytical approximations. The same is also true for negative chemical potentials, except for a small change of x_c as proposed by Hu and Silk (1993) (see Figg. 4b and 5b).

As already mentioned above the time evolution of μ is not exactly the same for the negative and positive case, particularly near z_{therm}. In the case $\mu = 0$ the numerical results yield much lower values of the thermalization redshift, as expected. Significant energy inputs are thermalized down to $z \simeq z_1$ (see Fig. 6) and leave very small tracks on the CMB spectrum.

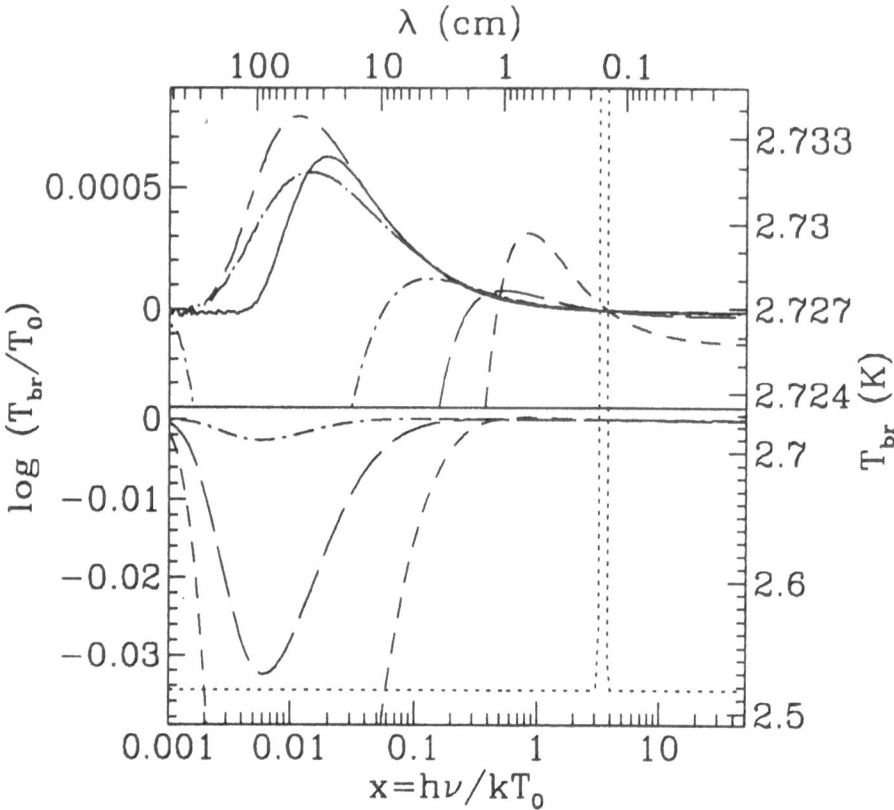

Fig. 6. Detailed evolution of a distorted spectrum in the very peculiar case $\mu = 0$. Although a conspicuous amount of photon is injected, $\Delta n_\gamma / n_i = 0.27$ with a large amount of energy $\Delta \epsilon / \epsilon_i \simeq 0.375$, (see eq. (28)), nonetheless a very small distortion remains at the recombination. The dotted line refers to the initial spectrum at $z_{dec} \simeq 4.33\,10^5$ corresponding to $y_e \simeq 10.6$. Bottom panel: at early times a large distortion in the RJ region appears that is subsequently strongly smoothed by the combinated action of Compton scattering, bremsstrahlung and radiative Compton (dashed line: $z \simeq 3.76\,10^5$, $y_e \simeq 7.95$; long dashes: $z \simeq 3.07\,10^5$, $y_e \simeq 5.29$; dots plus dashes: $z \simeq 2.18\,10^5$, $y_e \simeq 2.64$). Top panel: the spectrum at the above mentioned times together with that at lower redshifts (dots plus long dashes: $z \simeq 1.39\,10^5$, $y_e \simeq 1.06$; dashes plus long dashes: $z \simeq 9.67\,10^4$, $y_e \simeq 0.51$; solid line: $z = z_{rec}$, $y_e = 0$). We note that between $z \simeq 9.67\,10^4$ and z_{rec} the spectrum is stationary except at very long wavelengths, where the bremsstrahlung absorption operates; the relaxation is achieved in a time corresponding to $y_e \simeq 10$ as in the cases showed in Figg. 4 and 5 ($\Omega_b = 0.1$, $H_0 = 50$, $\Omega_T = 1$).

For decays producing positive chemical potentials, interesting limits on the energy density in the radiative channel $m_X n_X B_\gamma$ as function of the decay time τ have been derived by Hu and Silk (1992) and Burigana et al. (1993). The limits are in some region more stringent than those derived by stellar evolution (Raffelt et al. 1989).

The decay can be constrained also in the case of negative μ. For instance in the case of $x_X \leq 0.1$ and $z_{dec} \approx z_1$ the COBE/FIRAS 95% CL limit $\mu > -3 \times 10^{-4}$ implies

$$B_\gamma n_X / n_i \leq 2 \times 10^{-4}. \tag{29}$$

3.2 Vacuum decay

Many authors have suggested that vacuum energy density decays with time from substantial values in the early universe. Because of its negative pressure the vacuum behaves as energy source in the universe. The effects on nucleosynthesis of the vacuum decay constrain the vacuum energy density to be a small fraction of the total energy density of the universe (Freese et al. 1987). Bartlett and Silk (1990) showed that vacuum decay could distort significantly the CMB spectrum in the Wien as well in the Rayleigh–Jeans (RJ) region. While Bartlett and Silk (1990) have assumed that the vacuum energy decay as a heating source of the electrons, more recently Overduin et al. (1993) make the hypothesis that the decay at each time produces photons with BB spectral distribution. The resulting global spectrum has been compared to that of the CMB to bound vacuum energy density. Indeed in this way they have found constraints much weaker that those obtainable from a full discussion of the possible distortions.

Burigana et al. (1993) calculated the distortions by assuming the decay as a heating source of the electrons, as suggested by Bartlett and Silk (1990). However because the number of photons produced in the framework proposed by Overduin et al. (1993) is, at least at early epochs, negligible respect to those of the CMB, the results also hold for their case.

The vacuum energy density is assumed to be proportional to the energy density ϵ_k present in some form in the universe

$$\epsilon_v = \chi \epsilon_k. \tag{30}$$

The calculation by Burigana et al. (1993) showed that the limits on $\mu_0 \lesssim 1.0 \times 10^{-4}$ (Mather et al. 1993) entail $\chi \lesssim 1. \times 10^{-5}$. Thus the vacuum energy density was not a relevant fraction of the overall energy density. More tight constraints derive from the FIRAS limit to the comptonization distortions (see §4.1).

3.3 Damping of primordial density perturbations

The dissipation of primordial density perturbations is quite effective soon after the thermalization, producing BE–like distortions. Therefore limits on chemical potential set constraints on the dissipation rate (that in turn bound the initial power spectrum of density perturbations) that are independent and complementary (information on small scales) to those coming from CMB isotropy measurements (Sunyaev and Zeldovich 1970b; Daly 1991; Barrow and Coles 1991).

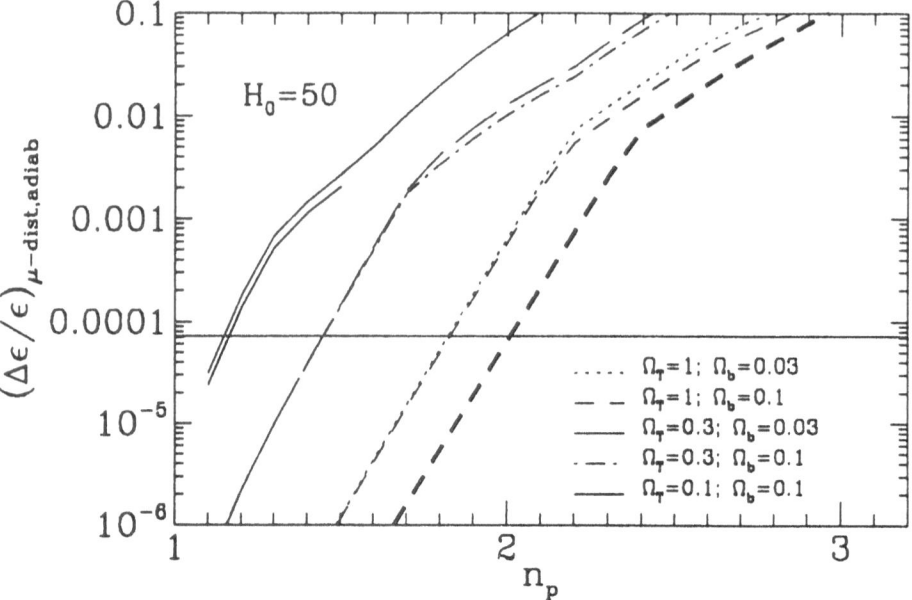

Fig. 7. Estimate of Bose–Einstein like distortion in terms of $\Delta\epsilon/\epsilon \simeq 0.7\mu$ in the case of adiabatic perturbations. The horizontal line shows the 2σ upper limit on $\Delta\epsilon/\epsilon$ based on recent FIRAS data (Mather et al. 1993). The calculations have been done neglecting photon production; dissipation is taken into account since a critical redshift z_c of order of $z_{therm}/3$. These semplifications introduce only very small changes in the final results as it can be seen from the case $\Omega_b = \Omega_T = 0.1$ for which the results of computations with no semplifications are also presented (short continuous line). No bias is assumed in all cases except for the heavy line, where $b = 3$ is considered to illustrate the effect of bias.

Following Daly's approach, Burigana et al. (1993) evaluated the distortions generated by the damping of primordial fluctuations due both to non-linear dissipation and to photon diffusion. In their computations photon production by bremsstrahlung and radiative Compton is included.

The spectrum of the primeval fluctuations was assumed to be a power law

$$|\delta_k|^2 = A k^{n_p}, \tag{31}$$

with such a normalization that yields rms matter fluctuations $\delta_{m,rms} \simeq 1$ on scale $\lambda \simeq 8h^{-1}\mathrm{Mpc}$ at the present epoch ($z = 0$).
In Fig. 7 the values of chemical potential produced at early epochs by the dissipation of adiabatic primordial density perturbations are presented as function of their spectral index n_p for various possible universes.
The constraints derived from the available data on the index of the initial power spectrum n_p are consistent with the COBE/DMR results $n_p = 1.1 \pm 0.5$ (1σ errors, Smoot et al. 1992).

The weak dependence of n_p on $\Delta\epsilon/\epsilon$ (apparent from Fig. 7) is easily understood on the basis of analytical approximations (Wright et al. 1993). Letting $r = log(\lambda_H/\lambda_{nl}(z_h))$ the ratio of the present horizon scale to the scale of non-linear dissipation at the effective redshift z_h, the bound on n_p reads

$$n_p \lesssim 1 + \frac{1}{r}\left(7 - 2log(\delta_{rH}/10^{-5}) + log(\Delta\epsilon/\epsilon/10^{-4})\right), \qquad (32)$$

where δ_{rH} is the radiation density fluctuation on very large (horizon) scales. The parameter r is a function of the cosmological model through the usual scaling of perturbation growth as well as the photon producing mechanisms. For instance $r \sim 7$ for the case $H_0 = 50$, $\Omega_T = 1$, $\Omega_b = 0.06$.
In some cases (e.g. $H_o = 50$, $\Omega_T = 0.1$, $\Omega_b = 0.1$) the constraints are quite close to challenge $n_p = 1$. On the other hand it should be noted that the dependence on Ω_T is such that in universes with $\Omega_T = 1$, $\Omega_b \leq 0.1$ and $n_p \leq 1.5$ dissipation of primordial perturbation would produce practically undetectable spectral distortions in the RJ region (Daly 1991; Barrow and Coles 1991; Burigana 1993).

4 Evolution of distorted spectra at $z \leq z_1$

At redshifts $z \leq z_1$ the kinetic equilibrium can no longer be maintained. As a consequence time and mechanism of energy injection leave more clear imprints on the CMB spectrum.
It has been shown that the non-equilibrium occupation number of photons (with no bound states) can be generally described by means of a linear superposition of Planck spectra (Zeldovich et al. 1972)

$$\eta(\nu, t) = \int R(T, t)\eta_{Pl}(\nu, T)dT, \qquad (33)$$

with a temperature distribution function $R(T, t)$ specific for each case.
In a number of cases spectra with small departures from the Planck form are rather accurately described by

$$\eta \simeq \eta_s(x) = \frac{1}{(4\pi u)^{1/2}}\int_0^\infty \frac{1}{\exp(x'/\phi_i) - 1} \times \exp\left\{-\frac{[\ln(x/x') + 3u]^2}{4u}\right\}\frac{dx'}{x'}, \qquad (34)$$

where $x = h\nu/kT_r$.
The parameter u is uniquely determined by the amount $\Delta\epsilon$ of extra energy injected in the radiation field; for $u \ll 1$, $u \simeq \frac{1}{4}(\Delta\epsilon/\epsilon_i)$ and $\phi_i = (T_i/T_r)(1 + z)/(1 + z_i) \simeq (1 + \Delta\epsilon/\epsilon_i)^{-1/4}(\simeq 1$ for very small distortions), being z_i the initial redshift of the heating process.

In the case that distortions are produced through heating of the matter, whatever the ratio T_e/T_r may be, the parameter u is given by

$$u(t) = \int\frac{\phi - \phi_i}{\phi}\frac{dt}{t_C} = \int\frac{\phi - \phi_i}{\phi}(kT_e/mc^2)n_e\sigma_T c\,dt. \qquad (35)$$

If $T_e \gg T_r$, then $u = y_e$. This is the case of scattering of cool photons off hot electrons and y_e is then related to the fractional amount of energy injected (in this context we will use the symbol y at place of y_e): in these circumstancies eq. (34) accurately describes the distorted spectrum even for large energy injections. Moreover Zeldovich et al. (1972) showed that the primeval small scale motions in the plasma can be described as random motions of plasma clouds with maxwellian velocity distribution and a planckian spectrum of radiation at temperature T_R. The corresponding temperature distribution function is

$$R(T, w) = \frac{1}{(4\pi w)^{1/2}} \exp\left[-\frac{1}{4w}\left(\frac{T}{T_R} - 1\right)^2\right], \qquad (36)$$

where $w = 1/6(v/c)^2$ (v is the mean squared velocity). It is interesting to notice that in the case of small distortions ($y \ll 1$, or $u \ll 1$, or $w \ll 1$ and $(T_e/T_R - 1) \ll 1$), the temperature distribution functions reduce to the same expression.

Chan and Jones (1975) pointed out that a very similar formula is also valid for the case of energy input deriving from primordial turbulence dissipation (in the hypothesis of small distortions).

The representation of the non–equilibrium radiation field by the above formulae allows for a straightforward derivation of important spectral properties: the brightness temperature in the RJ region is

$$T_{RJ} = \frac{h\nu}{k}\eta(\nu, T) = T_i\frac{1+z}{1+z_i}\exp(-2y) \simeq T_i\frac{1+z}{1+z_i}(1 - 2y) \simeq T_r(1 - 3y); \quad (37)$$

the fractional amount of energy injected in the radiation field is

$$\frac{d\epsilon}{dy} \simeq 4\epsilon; \qquad (38)$$

the stationary electron temperature (Peyraud 1968; Zeldovich and Levich 1970) is

$$T_{eq} \simeq T_i\frac{1+z}{1+z_i}(1 + 5.4y); \qquad (39)$$

finally in the low to moderate frequency region ($yx_i^2 \ll 1$ with $x_i = x/\phi_i$) the spectrum can be approximated by

$$\eta = \eta_{Pl}(x_i)\left[1 + \frac{yx_i\exp(x_i)}{\exp(x_i) - 1}\left(\frac{x_i}{\tanh(x_i/2)} - 4\right)\right]. \qquad (40)$$

As a consequence of eqq. (37) and (39) we expect in the RJ region a jump in the brightness temperature,

$$\frac{\Delta T}{T} \approx 7.4y, \qquad (41)$$

just around

$$\lambda_B = 94\left(\frac{T_0}{2.7\,K}\right)^{11/4}\left(\frac{H_0}{50}\right)^2\Omega_b^{-1}g_B(x_B)^{-1/2}z_h^{-1/4} \text{ cm}, \qquad (42)$$

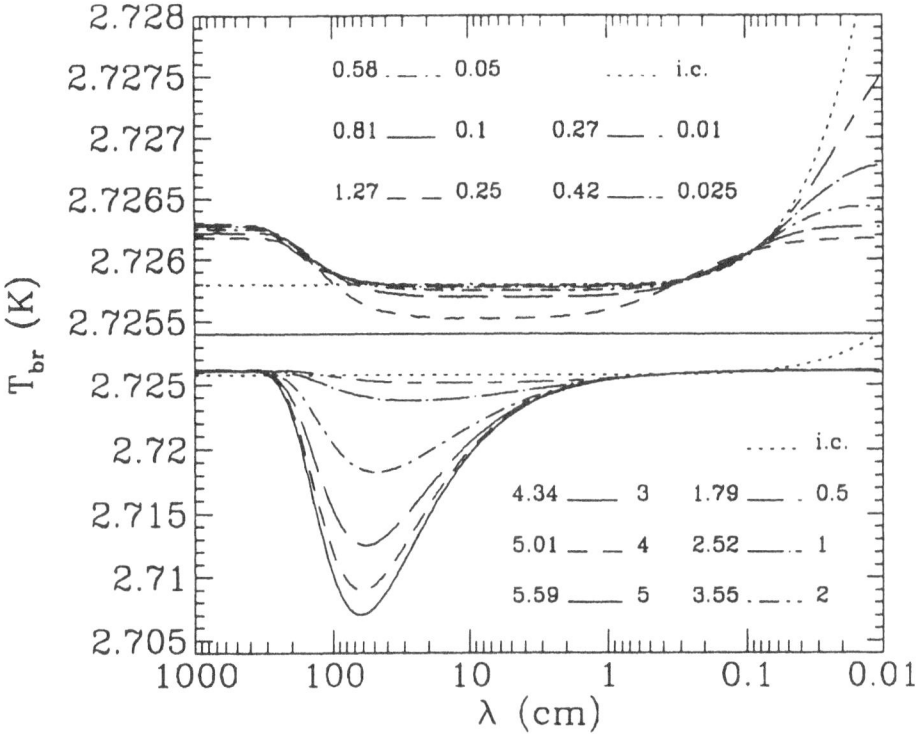

Fig. 8. Distorted spectra formed as a consequence of an energy injection $\Delta\epsilon/\epsilon_i = 10^{-4}$ occurring at redshifts z_h indicated on the left of each horizontal dash (in units of 10^5), corresponding to values of y_e indicated on the right of the dashes. The initial spectrum, assumed to be a superposition of black-bodies, is represented by the dotted lines (labeled i.c.). The bottom panel shows the results for injections occurring at larger redshifts ($H_0 = 50$, $\Omega_b = 0.03$, $T_0 = 2.726$K).

if $z_h \gg z_{eq}$, or around

$$\lambda_B = 1.3 \times 10^3 \left(\frac{H_0}{50}\right)^{-3/2} \Omega_T^{1/4} \Omega_b \left(\frac{T_0}{2.7\ K}\right)^{15/4} \left(\frac{T_e}{T_r}\right)^{7/4} g_B(x_B)^{-1/2} z_h^{-1/2}\ \mathrm{cm}\,,$$

$$(43)$$

if $z_h \ll z_{eq}$ (see Fig. 8, top panel).

The same relationships hold also in the case of CMB photons interactions with a non relativistic plasma with Thomson optical depth $\tau_T < 1$. Actually Sunyaev (1980) and Fabbri (1981) have demonstrated that the Kompaneets equation is still valid in the case of low probability scattering (such as in clusters of galaxies for the Sunyaev–Zeldovich effect) at least for frequencies $x < 10$.

Relativistic effects become important when a reheating of the baryonic matter result in a very high electron temperature, $T_e \gtrsim 50$ keV, (Wright 1979; Fabbri 1981), as required for instance by models of diffuse bremsstrahlung for the X–

ray Background Radiation (XBR) suggested in the past (Field and Perrenod 1977). In this circumstances the effect on CMB spectrum must be evaluated by computing the frequency shift distribution averaged over the scattering angles, with the proper velocity distribution and convolving with the initial spectrum.

By fitting FIRAS data with the above approximate equation (40) for comptonized spectra Mather et al. (1993) found (1σ errors)

$$y = (3 \pm 11) \times 10^{-6}. \tag{44}$$

However, as already noticed, distortions after z_1 keep some memory of the time and mechanism of heating (see Fig. 8).

As expected for $y_e(z_h) > 1$ (Fig. 8, bottom panel) the trend is toward a BE like spectrum; for lower $y_e(z_h)$ the minumum at long wavelengths tend to broaden, approaching the value corresponding to a comptonized spectrum given by eq. (37). On the other hand at shorter wavelengths for $y_e(z_h) \gtrsim 0.1$ BE–like spectra form, even if we start with comptonized spectra (Zeldovich and Sunyaev 1969; Illarionov and Sunyaev 1974; Burigana et al. 1991b). Therefore the classical approximation to comptonized spectra (eq. (40)) strictly holds only for $y_e(z_h) \lesssim 0.05$. As a consequence the mentioned FIRAS limits are adequate for Comptonization at low redshifts $z_h \lesssim 3 \times 10^4$.

Fig. 8 shows that with $0.1 < y_e(z_h) < 0.5$ the amplitude of the distortions in terms of $\Delta T/T$ is small at all wavelengths. This may imply less stringent limits on heating processes occurring at the corresponding cosmological time.

4.1 Constraints on physical processes at $z \leq z_1$

Particle decays might significantly distort the CMB spectrum at $z \leq z_1$. An approximate relationship of the Compton parameter u with the injected energy and photon number densities can be derived in a way analogous to that used for the chemical potential in §3.1 in the limit $u \ll 1$

$$u \simeq \frac{1}{4}\left(\frac{1 + \Delta\epsilon/\epsilon_i}{(1 + \Delta n_\gamma/n_i)^{\frac{4}{3}}} - 1\right) = \frac{1}{4}\left(\frac{1 + R_X B_\gamma x_X/\bar{x}_{CMB}}{(1 + R_X B_\gamma)^{\frac{4}{3}}} - 1\right). \tag{45}$$

Fig. 9 shows that the numerical calculations in the cases of energy injection by particle decays predict final spectra similar to those obtained with different heating mechanisms, provided that the value of u and $y_e(z_h)$ is the same. Comparison of very peculiar cases with $u = 0$ (see Fig. 10) to cases with the same injected energy (Fig. 9, bottom panel) shows that a non-negligible number of photons helps in decreasing the distortions at least for $y_e(z_h) \gtrsim 0.1$. It is worth noticing that in the decaying particles case comptonized spectra practically never form, because at late time enough to produce a comptonized spectrum ($y_e(z_h) \lesssim 0.05$) an excess of energetic photons not yet reprocessed appears.

In conclusion particle decay processes at redshifts z_h such that $y_e(z_h) \lesssim 2$ modify the spectrum in a way that keeps memory of the decay time, of the injected

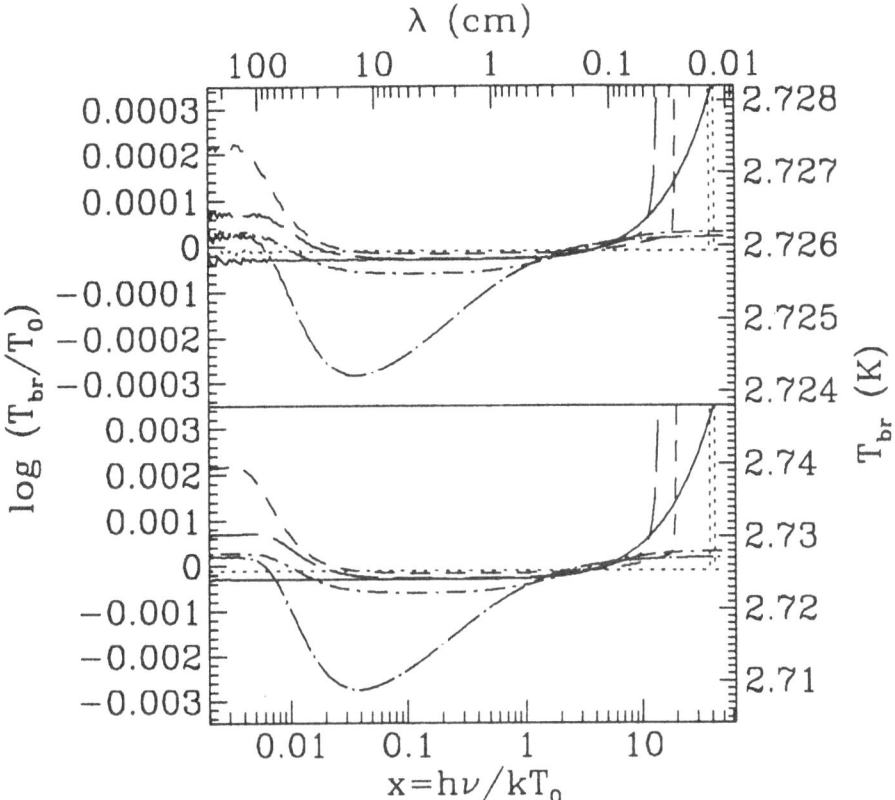

Fig. 9. Distorted spectra formed as a consequence of a radiative decay producing a "line" over an unperturbed black–body (dotted lines) and occurring at different cosmic times corresponding to $y_e = 0.01$ (dashed lines), 0.05 (long dashes), 0.25 (dots plus dashes) and 1 (dots plus long dashes) ($H_0 = 50$, $\Omega_T = 1$, $\Omega_b = 0.1$, $T_0 = 2.726$). Bottom panel: $\Delta\epsilon/\epsilon_i = 10^{-3}$ and $\Delta n_\gamma/n_i \simeq 6.75\ 10^{-5}$ so that $u = 2.275\ 10^{-4}$ (see eq. (45)). Top panel: $\Delta\epsilon/\epsilon_i = 10^{-4}$ and $\Delta n_\gamma/n_i \simeq 6.75\ 10^{-6}$ so that $u = 2.275\ 10^{-5}$. In the two cases $x_X \simeq 40$. For $y_e < 0.05$ a comptonized spectrum is not produced because there is no time enough for reprocessing the energetic photons, whereas for values of y_e greater 0.25 a Wien tail appears and the hollow in the RJ region increases with y_e. A comptonized spectrum (eq. 40) with the same value of comptonization parameter used in each panel is showed for comparison (solid line).

energy and photons. Therefore limits to these processes by the standard comptonized spectra (eq. (40) should be taken as tentative.

In the case of vacuum decay the electron temperature becomes very high ($T_e \gg T_r$) at low redshift ($z \approx 10^3$), as showed by Bartlett and Silk (1990), and the study is leaded back to the case of comptonization by hot electrons. From Fig. 11 it is apparent that the vacuum energy had to be only a tiny fraction $\chi \lesssim 10^{-6}$ of the important forms of energy density (see also §3.2).

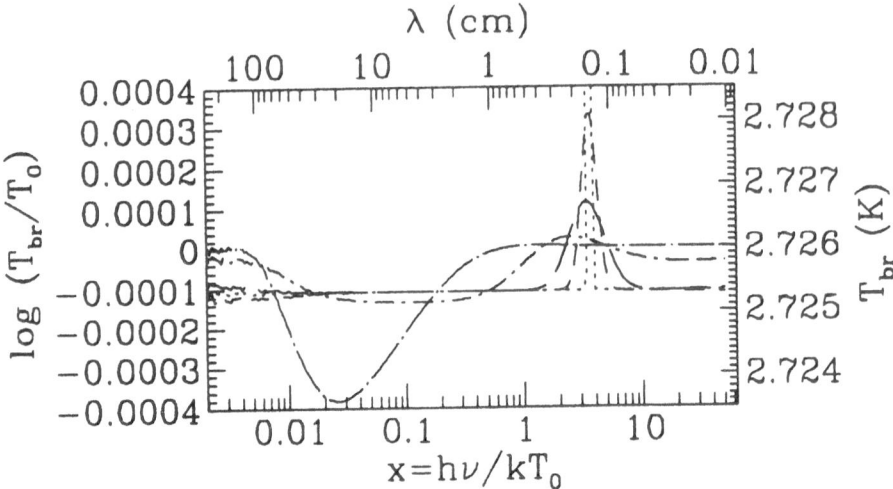

Fig. 10. Same as Fig. 9 but for the very peculiar case $u = 0$. Here $\Delta\epsilon/\epsilon_i = 10^{-3}$, as in bottom panel of Fig. 9, and consequently x_X results to be $\simeq 3.6$. As expected the non–negligible number of photons helps in decreasing the distortion. In spite of the "significant" amount of energy (i.e. potentially well detectable by FIRAS for processes injecting energy without photon production) for peculiar values of $y_e(z_{dec})$ the distortion is presently undetectable.

The comptonization distortions expected in the case of damping of adiabatic perturbations are rather small and set bounds on the primordial index of power spectrum n_p weaker than those derived from the limits on BE–like distortions. On the other hand in the case of isocurvature perturbations only comptonization distortions may survive, because the BE distortions are erased by the very large photon production rate via bremsstrahlung, due to the high value of matter density contrast δ_m (Daly 1991; Burigana 1993). For reasonable values of the small scale cut–off λ_m of the power spectrum of matter fluctuations and for a universe with $\Omega_T = \Omega_b = 0.1$ it was found $n_p \lesssim -0.2$ (Burigana et al. 1993).

4.2 Distortions generated during the recombination

As the expansion of the universe goes on, protons and electrons combine. Spectral distortions are also generated during and after the recombination, depending on the thermal and ionization history of the universe.

The standard theory of the recombination (Peebles 1968; Zeldovich et al. 1968) shows that distortions should be located at $\lambda \lesssim 300$ μm and at level $(\nu I_\nu \lesssim 10^{-8} \, erg/cm^2/s/sr)$ well away from any possible detection. Actually the expected sensitivity of DIRBE in this band is at least a factor of ten above the previously quoted limit. On the other hand Lyubarsky and Sunyaev (1983)

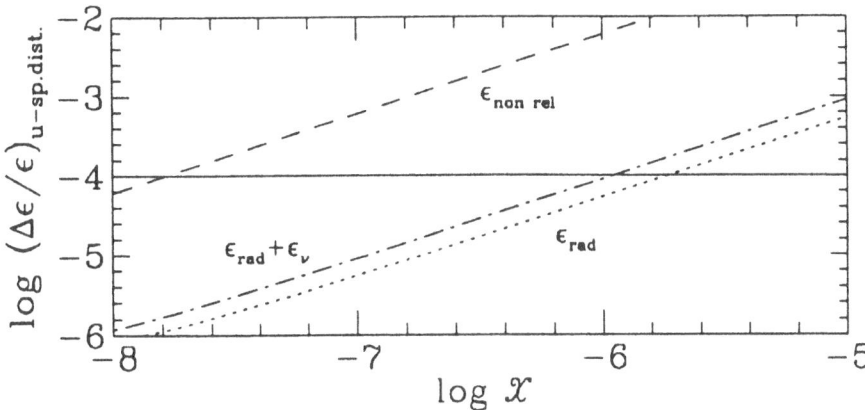

Fig. 11. Estimates of comptonization distortions produced by vacuum decay, for three choices of ϵ_k. If the vacuum energy density is assumed to be proportional to radiation or to radiation plus neutrinos energy density, the result is essentially indipendent of the thermal history after the recombination. In the case $\epsilon_k = \epsilon_{non\ rel}$ it is assumed that the universe remains ionized at low redshifts; on the other hand, if recombination occurs in the standard way, the results are essentially the same as in the other two cases. A cosmological model with $\Omega_T = 1$, $\Omega_b = 0.03$ and $H_0 = 50$ is assumed. The upper limit on comptonization distortion from FIRAS data is represented by the horizontal solid line.

showed that these distortions are enhanced in the submillimetric region by possible preexisting BE like distortions. However for values of the chemical potential μ within the limits imposed by CMB observations and for baryon densities consistent with standard primordial nucleosynthesis the expected distortions are at least a factor of ten smaller than the currently available limits.

4.3 Post–recombination distortions

Comptonization of CMB photons as well as bremsstrahlung emission occur in the intergalactic medium (IGM) if the ionization keeps at significant levels, distorting the CMB spectrum.

On the other hand Bartlett and Stebbins (1989) have argued that spectral observations of the CMB will ever require recombination in flat universes with standard nucleosynthesis. Of course this conclusion depends on the thermal history of the IGM.

To illustrate this point we have computed the distortions generated by free-free emission of an ionized IGM with thermal history compatible with the limits on the comptonization parameter. We present a case (Fig. 12) with $T_e \simeq 5 \times 10^3 > T_r$ and $u \simeq 2.36 \times 10^{-5}$, resulting in an excess of brightness temperature T_{br} at long wavelengths. A different case is presented in Fig. 13 with $T_e/T_r \simeq 0.2$

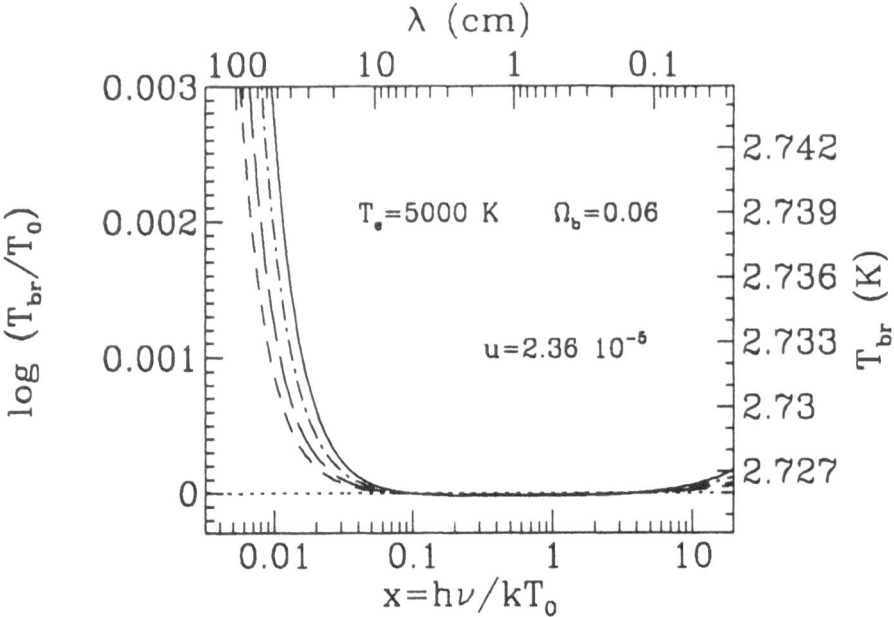

Fig. 12. Evolution of a BB spectrum since $z = 1000$ (dotted line) for the case of ionized matter at constant temperature $T_e = 5000$K. The spectrum at several times is showed: $z = 610$ (dashed line), 499 (long dashes), 245 (dots plus dashes) and present time (solid line) ($H_0 = 50$, $\Omega_T = 1$). The distorted spectrum is characterized by positive values of u and y_B as a consequence of the assumption $\phi > \phi_i$ (see eqq. (35) and (46)).

and $u \simeq -2.17 \times 10^{-5}$ wherein an important decrease of T_{br} is produced at long wavelengths by bremsstrahlung absorption.

It is interesting to notice that the expected distortion in the case of $T_e > T_r$ is much smaller than that expected in the other case. Indeed the comptonization parameter u is proportional to the kinetic energy density of the IGM $\epsilon_{IGM} \simeq 3n_e k T_e$, while for a homogeneous IGM the correspondent of the comptonization parameter u for the bremsstrahlung emission is

$$y_B = \int \frac{\phi - \phi_i}{\phi} t_{exp} K_B \phi^3 g_B dt \qquad (46)$$

with the Gaunt factor g_B weakly dependent on frequency. The two parameters may be positive or negative according to the sign of $(1 - \phi_i/\phi)$.

The two cases predict much larger distortions for the case with $T_e < T_r$, because of the different dependence of comptonization parameter u and y_B on the ratio $\phi = T_e/T_r$.

FIRAS data already constrain the comptonization parameter u so severely that in the case $T_e/T_r > 1$ searches for IGM emission would require measurements with few mK precision at long wavelengths, where foregrounds are increasingly

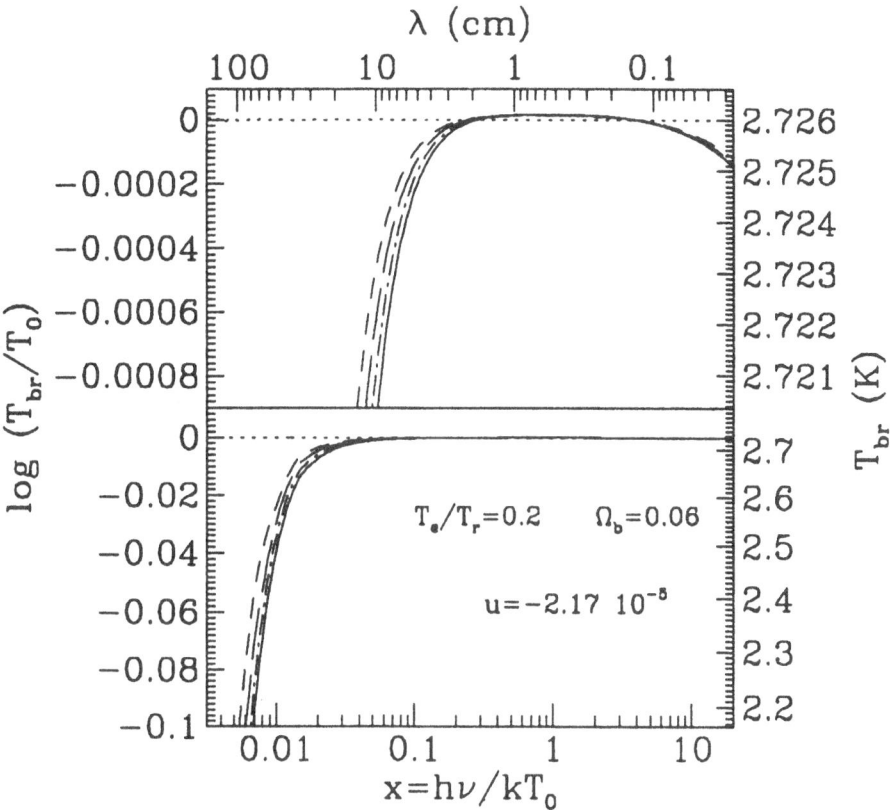

Fig. 13. Evolution of a BB spectrum since $z = 1500$ (dotted line) for the case of ionised matter with a constant ratio $T_e/T_r = 0.2$ between the matter and radiation temperature. The spectrum at several times is showed: $z = 749$ (dashed line), 499 (long dashes), 245 (dots plus dashes) and present time (solid line) ($H_0 = 50$, $\Omega_T = 1$). The distorted spectrum is characterised by negative values of u and y_B as a consequence of the assumption on the ratio T_e/T_r (see eqq. (35) and (46)). Of course at very long wavelengths, where bremsstrahlung is very efficient, the spectrum approaches to that of a black-body with temperature $T = T_e$. The top panel is only a blow-up of a part of the bottom one for sake of comparison between the distortions at submillimetric and RJ spectral regions.

important (see Bersanelli et al. these proceedings). On the other hand in the case $T_e/T_r < 1$ a thermal history compatible with FIRAS limits might produce distortions at level that hopely could be detectable in the next future by long wavelength experiments. As for the cooling mechanism, it has been suggested by Stebbins and Silk (1986) that recombinative cooling could yield $T_e \gtrsim 0.2T_r$.

A diffuse origin of the hard XRB ($E \geq 3$ keV) is largely ruled out by the present COBE/FIRAS data, because it would require IGM energy density $\epsilon_{IGM}(z = 0) \approx 10^{-13}\,\mathrm{erg\,cm^{-3}}$ and $y \approx 10^{-2}$ (Field and Perrenod 1977).

As argued by many authors, conventional sources are expected to inject energy in the IGM. Star formation in galaxies and AGNs are unable to transfer to the IGM an energy density $\epsilon_{IGM} \gtrsim 10^{-16}\,\mathrm{erg\,cm^{-3}}$; as a consequence $y \lesssim 10^{-5}$ is expected (De Zotti and Burigana 1992).

Formation of large–scale structure could also heat the IGM. Detailed calculations (Cen et al. 1990; Cen and Ostriker 1992; Cavaliere et al. 1991) have shown that the comptonization should range from few$\times 10^{-4}$ to few$\times 10^{-6}$. The most extreme cases of galaxy formation such as explosion driven formation are excluded by FIRAS data.

The results of the Gunn-Peterson test implies that the IGM is highly ionized at least from $z \approx 5$; however the associated comptonization is expected to be small $y \approx 10^{-7}$.

In conclusion FIRAS bounds on comptonization $y \lesssim 2.5 \times 10^{-5}$ exclude some scenarios of IGM evolution and point to a tepid or cold thermal history.

We gratefully acknowledge helpful and numberless discussions with G. De Zotti, A. Franceschini and L. Toffolatti.

References

1. Barrow, J.D., and Coles, P., 1991, *M.N.R.A.S.*,248, 52
2. Bartlett, J.G., and Silk, J., 1989, *Workshop on Particle Astrophysics: Forefront Experimental Issues*, Berkeley, CA (USA), E.B. Norman ed.
3. Bartlett, J.G., and Stebbins, A., 1991, *Ap. J.*,371, 8
4. Bersanelli, M., Smoot, G.F., Bensadoun, M., De Amici, G., and Limon, M., 1993, *contribution to this Workshop*
5. Burigana, C., 1989, *Thesis*, University of Padua
6. Burigana, C., 1993, *Ph. D. Thesis*, University of Padua
7. Burigana, C., Danese, L., and De Zotti, G., 1991a *Astr. Ap.*,246, 59
8. Burigana, C., De Zotti, G., and Danese, L., 1991b *Ap. J.*,379, 1
9. Burigana, C., et al., 1993, in preparation
10. Cavaliere, A., Menci, N., and Setti, G., 1991, *Astr. Ap.*,245, L21
11. Chan, K.L., and Jones, B.J.T., 1975, *Ap. J.*,200, 454
12. Cen, R.Y., Jameson, A., Liu, F., and Ostriker, J.P., 1990, *Ap. J.*,362, L41
13. Cen, R.Y., and Ostriker, J., 1992 *Ap. J.*,393, 22
14. Daly, R.A., 1991, *Ap. J.*,371, 14
15. Danese, L., and De Zotti, G., 1977, *Rivista Nuovo Cimento*, 7, 277
16. Danese, L.,and De Zotti, G., 1980, *Astr. Ap.*,84, 364
17. De Zotti, G., 1986, *Progress in Particle and Nuclear Physics*, Vol. 17, 117, ed. Faessler, A.
18. De Zotti, G., and Burigana, C., 1992, In *Highlights of Astronomy*, Vol. 9, 265
19. Fabbri, R., 1981, *Ap. Space Sci.*, 77, 529

20. Field, G.B., and Perrenod, S.C., 1977, *Ap. J.*, 215, 717
21. Freese, K., Adams, F.C., Frieman, J.A., and Mottola, E., 1987, *Nucl. Phys.* *B287*, 797
22. Gould, R.J., 1984, *Ap. J.*, 285, 275
23. Hu, W., and Silk. J., 1992, *Phys. Rev. Lett.*, submitted
24. Hu, W., and Silk. J., 1993, *Phys. Rev. D*, submitted
25. Illarionov, A.F., and Sunyaev, R.A., 1974, *Astron. Zh.*, 51, 1162 [*Sov. Astr.*, 18, 691 (1975)]
26. Kawasaki, M., and Sato, K., 1987, *Pubbl. Astron. Soc. Japan*, 39, 837
27. Kompaneets, A.S., 1956, *Zh. E. T. F.* 31, 876 [*Sov. Phys. JETP* 4, 730 (1957)]
28. Lyubarsky, Yu.E., and Sunyaev, R.A., 1983, *Astr. Ap.*, 123, 171
29. Mather, J.C. et al., 1993, *Ap. J.*, in press
30. Overduin, J.M., Wesson, P.S., and Bowyer, S., 1993, *Ap. J.*, 404, 1
31. Peebles, P.J.E., 1968, *Ap. J.*, 153, 1
32. Peyraud, J., 1968, *J. Physique*, 29, 306
33. Raffaelt, G., Dearborn, D., and Silk, J., 1989, *Ap. J.*, 336, 61
34. Rybicki, G.B., Lightman, A.P., 1979, *Radiative processes in astrophysics*, Wiley, New York
35. Smoot, G.F. et al. 1992 *Ap. J.*, 396, L1
36. Stebbins, A., and Silk, J., 1986, *Ap. J.*, 269, 1
37. Sunyaev, R.A., 1980, *Pis'ma Astr. Zh.*, 6, 387 [*Sov. Astr. Letters*, 6, 213]
38. Sunyaev, R.A., and Zeldovich, Ya.B., 1970a, *Ap. Space Sci.*, 7, 20
39. Sunyaev, R.A., and Zeldovich, Ya.B., 1970b, *Ap. Space Sci.*, 9, 368
40. Weymann, R., 1966, *Ap. J.*, 145, 560
41. Wright, E.L., 1979, *Ap. J.*, 232, 348
42. Wright, E.L. et al., 1993, *Ap. J.*, in press
43. Zeldovich, Ya. B., Illarionov, A.F., and Sunyaev, R.A., 1972, *Zh. Eksp. Teor. Fiz.*, 62, 1216 [*Soviet Phys. JETP*, 35, 643]
44. Zeldovich, Ya. B., and Levich, E.V., 1968, *Zh. Eksp. Teor. Fiz.* 55, 2423 [*Soviet Phys. JETP*, 28, 1287 (1969)]
45. Zeldovich, Ya. B., Kurt, V.G., and Sunyaev, R.A., 1968, *Zh. Eksp. Teor. Fiz.* 55, 278 [*Soviet Phys. JETP*, 28, 146]
46. Zeldovich, Ya. B., and Sunyaev, R.A., 1969, *Ap. Space Sci.*, 4, 301

Medium Scale CBR Anisotropy Measurements: UCSB South Pole HEMT (1990-91) and MAX 3 (1991)

Peter Meinhold[1,2] with the ACME-HEMT and MAX Collaborations

[1] The University of California at Santa Barbara
[2] The NSF Center for Particle Astrophysics

1 Introduction

The COBE detection of anisotropy in the Cosmic Background Radiation has spurred interest in comparisons between large, 'medium' and small angular scales. Since the COBE announcement, most calculations have used theoretical scenarios of Galaxy formation, along with a normalization of the primordial perturbation spectrum derived from the COBE data, to predict fluctuations at angular scales of 0.5- several degrees. A number of experiments have been done at these angular scales, with interesting results. CBR anisotropy results from the UCSB South Pole expedition (1990-1991), and the third flight of the Millimeter Anisotropy Experiment (MAX) experiment (1991) are presented. These two experiments probe the angular range from roughly 10 arcminutes to 3 degrees, crucial for testing theories of structure formation.

Our HEMT measurements made from the South Pole over a range of frequency from 25 to 35 GHz, with a beamsize of 1.5 degrees, have been used to set stringent constraints on structure formation theories normalized by the COBE detection. For a Gaussian Autocorrelation Function (GACF) model, these data provide an upper limit of 1.5×10^{-5} at 1.2 degrees. Structure is observed in the data at a level of $\Delta T/T \approx 8 \times 10^{-6}$, with spectrum consistent with that expected from CBR fluctuations, but more consistent with Galactic foreground contamination.

Balloon borne bolometric measurements by the MAX experiment in the frequency range from 180 to 360 GHz have detected interstellar dust, as well as significant structure possibly due to CBR fluctuations. Two separate regions of sky were measured, and after subtraction of an interstellar dust (ISD) component, the data from one region have been used to derive a new upper limit for a GACF of $\Delta T/T < 2.5 \times 10^{-5}$ at 25 arcminutes. The second region (with no ISD removal needed) shows a highly significant detection of structure, with a spectrum consistent with CBR anisotropy. The observed structure corresponds to a GACF amplitude of $\Delta T/T = 4.2^{+1.7}_{-1.1} \times 10^{-5}$, for a coherence angle of 25 arcminutes.

It should be explicitly stated here that the GACF upper limits are only quoted in order to allow a relatively experiment independent way to compare results. GACF limits at a given angular scale can be compared with each other, and can be used to test the (common) assumption of Gaussian *distributed* fluctuations as opposed to fluctuations well approximated by a GACF.

2 Foreground Contamination

Experiments attempting to measure CBR anisotropy at the level of a few parts in 10^6 are prone to systematic errors associated with four principal foregrounds: Emission from the Earth's atmosphere, off-axis (sidelobe) pickup of emission from the Earth, Moon, Sun, or other objects, diffuse emission from the interstellar medium (warm dust or radiation from charged particles), and extragalactic radio sources.

2.1 Atmosphere

The two experiments described here, ACME-HEMT 1990-1991, and the MAX experiment, rely on a choice of site and operating frequencies to reduce atmospheric emission. The ACME-HEMT system is a coherent amplifier system, and uses relatively narrow passbands (4 bands covering a total of 10 GHz centered at 30 GHz) in a frequency region with low atmospheric emission, operating from a high dry site (the South Pole). MAX is a bolometric system, operating over a much broader frequency span (3 bands at roughly 20-30 % bandwidth centered near 180, 270 and 360 GHz), and thus receives emission from many more atmospheric lines. MAX consequently observes from an altitude of 35 km to reduce the atmospheric signal.

2.2 Sidelobe Contamination

Both experiments use a Gregorian telescope with 1 meter focal length, underfilling both the elliptical secondary and the primary mirrors. The primary mirror and associated telescope mount and pointing system are common to the two experiments, although configured differently for ground and balloon operation. The underfilled optics and lack of obstruction in the primary beam afford very low sidelobe response for each system. Large reflective baffles are used to further reduce the effects of far sidelobe response. The telescope response has been mapped in one dimension to a level of more than 65 dB below the on-axis response for both ACME-HEMT and MAX. Measurements made during the third flight of MAX imply no significant contamination from sidelobe response. Data taken closer than 65 degrees from the Sun during the ACME-HEMT South Pole 1990-1991 expedition are possibly contaminated by solar emission at the part in 10^5 level, and consequently these data are rejected.

2.3 Galactic Foregrounds

Figure 1 shows the spectral characteristics of the expected Galactic foregrounds, in addition to that of a CBR anisotropy. The bandpasses of the two instruments are sketched in to illustrate the spectral discrimination used in the experiments. The absolute levels of the three Galactic foregrounds have been arbitrarily chosen, but are consistent with the levels expected for reasonably good regions away from the Galactic plane. Note that ACME-HEMT is best able to distinguish between synchrotron and bremsstrahlung radiation and CBR anisotropy, while MAX is extremely good at distinguishing ISD emission from CBR.

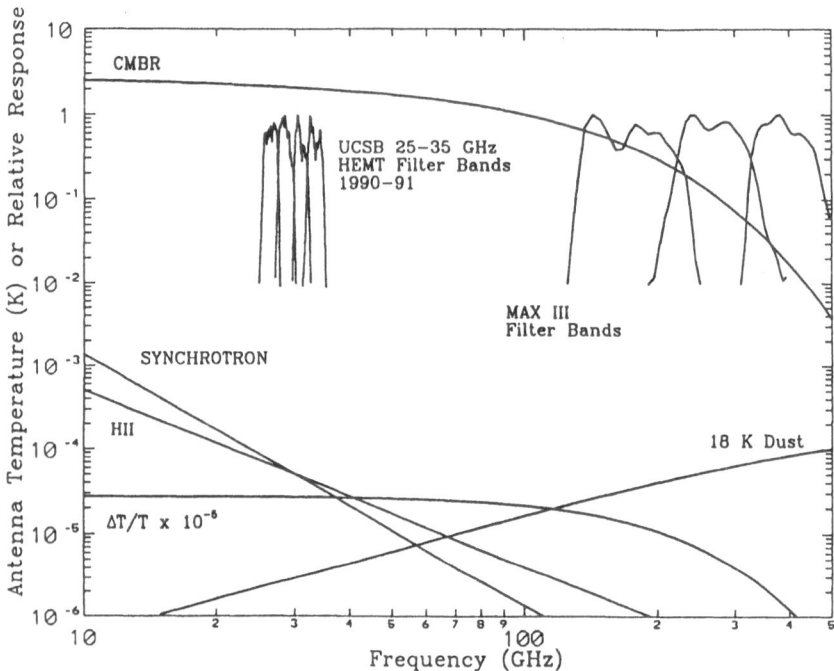

Fig. 1. Spectral characteristics of Galactic foregrounds far from the Galactic plane, presented in terms of antenna temperature. Power received in a single mode HEMT system is proportional to T_A, while that received by a constant throughput system like MAX is proportional to $\nu^2 T_A$. Free-free and Synchrotron specific intensity are set equal to 50 μK at 30 GHz, assumed to scale as $S(\nu) \propto \nu^{-2.1}$ and $S(\nu) \propto \nu^{-2.7}$ respectively. Dust emission assumes 18.5 K dust with emissivity $\propto \nu^{1.4}$

2.4 Extragalactic Sources

Flat spectrum extragalactic radio sources pose similar problems for ACME-HEMT and for MAX. Many types of spectra have been observed from different objects, making discrimination of CBR fluctuations from spectral information alone problematic. The general technique is to use low frequency surveys at high resolution, and morphological considerations to identify possible extragalactic contaminating sources. Potentially contaminated regions can then be mapped using ground based high resolution telescopes at several frequencies.

3 ACME-HEMT South Pole, 1990-1991

ACME-HEMT 1990-1991 refers to an expedition made to the South Pole from November, 1990 to January, 1991 to measure CBR anisotropy using the ACME telescope equipped with an extremely low noise, liquid ^{4}He cooled, direct amplification High Electron Mobility Transistor (HEMT) receiver (Pospieszalski et al, 1990). The telescope produced a 1.5 degree full width at half maximum (FWHM) response at 30 GHz, moved sinusoidally on the sky with peak to peak separation of 2.1 degrees. The beamsize varies as FWHM $\propto \nu^{-1}$. Data from this expedition have recently been published in Gaier et al (1992), and Schuster et al (1993).

3.1 Detector

The detector consisted of a HEMT amplifier cooled to 4.2 K with liquid Helium, with a noise temperature of 30 K, and a 10 GHz bandpass centered around 30 GHz. This was followed by a warm amplifier and a set of circulators and filters designed to produce 4 channels of 2.5 GHz each. In the absence of atmospheric noise, the expected sensitivity was 1.4mK$\sqrt{\text{sec}}$ in terms of $\Delta T/T$ for each channel. The actual measured noise (during good weather) varied from 1.8 to 3 mK $\sqrt{\text{sec}}$.

3.2 Measurement Strategy and Raw Data Set

The telescope beam was chopped through 2.1 degrees peak to peak, at 8 Hz. A lockin amplifier demodulated the detector output using a square wave weighting phased with the position of the secondary mirror, providing a first difference measurement of the sky on 0.5 second timescales. The basic measurement strategy consisted of sequentially stepping through a set of points, integrating on each for roughly 20 seconds. The telescope then reversed direction and stepped back through the points. The points were chosen to make the positive lobe from one integration overlay the negative lobe from the adjacent one. This motion over all the points and back is referred to as a full scan. A set of data was obtained including 24 hours worth of 9 point full scans on each of 6 adjacent elevations, as well as deeper integrations on a 13 and a 15 point strip overlaying some of the 9 point strips. The region covered is shown schematically in Figure 2. Measurements of the Moon, the Sun, the Galactic plane, and the Large Magellanic Cloud (LMC) were also performed for pointing and telescope calibration. The detector was calibrated with an ambient load 1 to 2 times per day.

The raw data were edited according to weather, (bad weather were usually identified by a dramatic increase in the noise), calibrations, and chopper instabilities. The data were fit to remove slowly varying offsets and a time varying gradient due to large scale atmospheric structure. The effects of this fitting procedure have been included in the calculations comparing the data to models. The data were then binned in angle and averaged. An error bar was calculated for each bin from the dispersion of the data in the bin, assuming Gaussian noise.

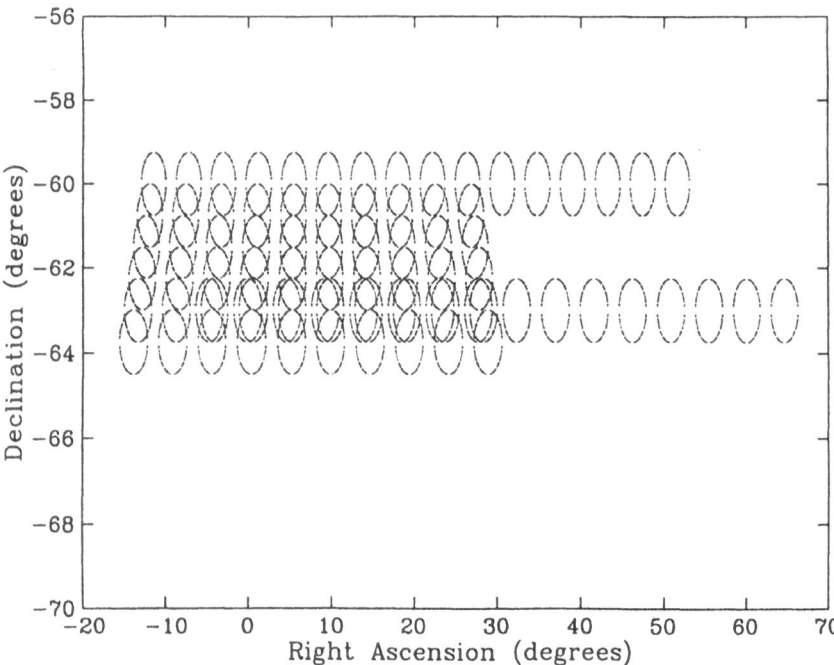

Fig. 2. Orientation of ACME-HEMT 1990-1991 data set. Circles represent half power points of positive or negative lobes of telescope response.

Typical per channel error bars for the best strips are about 22 μK, with a lowest error of 18 μK.

3.3 9 Point Data Set

The 9 point data set with the most integration time has been published in Gaier et al (1992). It consisted of 9 adjacent points centered near $\alpha = 0.5$ hours, $\delta = -62.25°$. This data set is shown in Figure 3. There is a signal in the three lower frequency channels which is essentially absent in the highest frequency channel, suggesting to the eye at least that the signal is not thermal (as CBR would be). The analysis of this data set has been performed a number of different ways by ourselves as well as others. The most probable signal is spectrally inconsistent with CBR anisotropy at the 2 sigma (95 % confidence) level, a result obtained via a number of techniques. The four channels are however *marginally* consistent with a 20 μK rms CBR anisotropy. The effects of the beamsize variation between channels have not been included in the analysis. Inclusion of this variation in the spectral calculation tends to reduce the probability that the measured structure could be CBR anisotropy. Results from our published analysis of this data set, using the highest frequency channel for setting upper limits to CBR anisotropy, are included in Figure 9. For that calculation we have assumed that the signal in the lower frequency channels is due to foreground contamination and that the highest frequency channel is thus the least contaminated. This upper limit is increased by about 25 % if one assumes a 20 μK rms CBR anisotropy, the amplitude implied by coadding the 4 channels assuming a CBR spectrum.

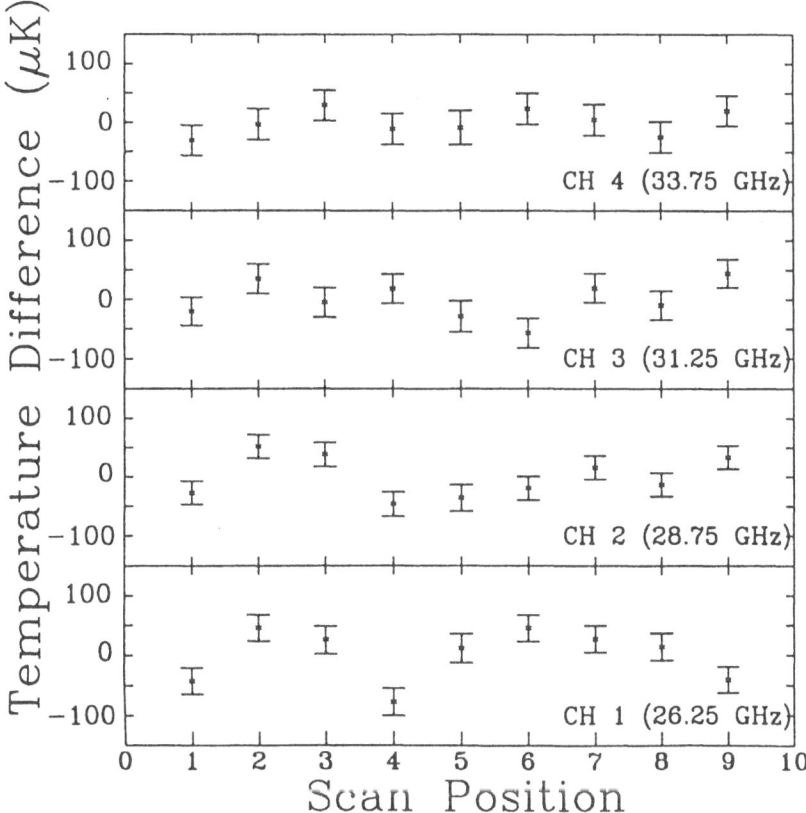

Fig. 3. 9 point data set from $\delta = -62.25°$. Vertical axis is in antenna temperature units, which are the same as thermodynamic temperature units for these frequencies.

3.4 15 Point Data Set

The deepest integration from this expedition was performed on 15 points centered near $\alpha = 2$ hours, $\delta = -63°$. See Schuster et al (1993). A portion of these data were taken with the Sun substantially closer in angle to the measurement region. Some evidence for solar contamination prompted the removal of all data when the Sun was closer than 65 degrees from the measured point. This resulted in only 13 valid data points, which are displayed in Figure 4. These data show a signal, present in all four channels, with substantially higher signal to noise than the signal seen in Figure 3. The spectrum of the most probable signal is consistent with CBR, but is more consistent with Galactic emission. Figure 5 shows the results of coadding the separate channels, under the assumption that all signal is due to CBR anisotropy (this is not the most probable spectrum, but is useful for conservative upper limit calculations). Note that the error bars on this data set are about 11 μK or $\Delta T/T = 4 \times 10^{-6}$

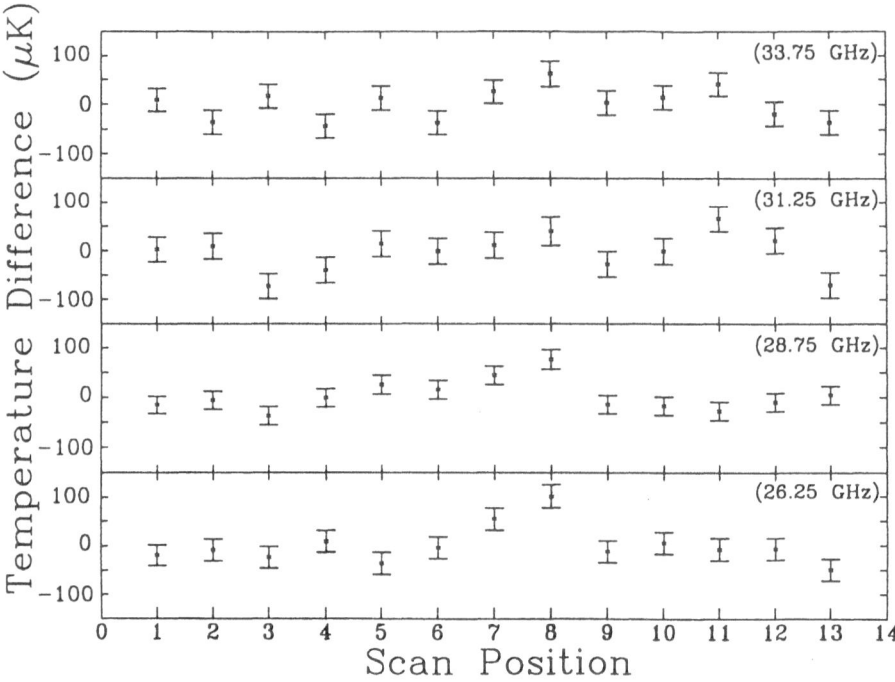

Fig. 4. 13 points of the 15 point data set from $\delta = -63°$.

A conservative analysis for upper limit calculations has been performed, assuming all the signal to be due to CBR anisotropy. Although the per pixel error bars for this data set are smaller than for the data described above, the presence of a signal results in an upper limit of $\Delta T/T < 1.6 \times 10^{-5}$, slightly above those obtained from the 9 point data set. The results are consistent with the 9 point data set, and would imply a CBR anisotropy of $\Delta T/T = 8 \times 10^{-6}$ for a GACF in the most sensitive angular range of 0.75 to 1.5 degrees if all the signal were attributed to the CBR. The upper limit curves from both data sets along with the most probable amplitude curve for the 15 point data set are included in the final plot, Figure 9. The consistency between the two data sets is clear from the figure.

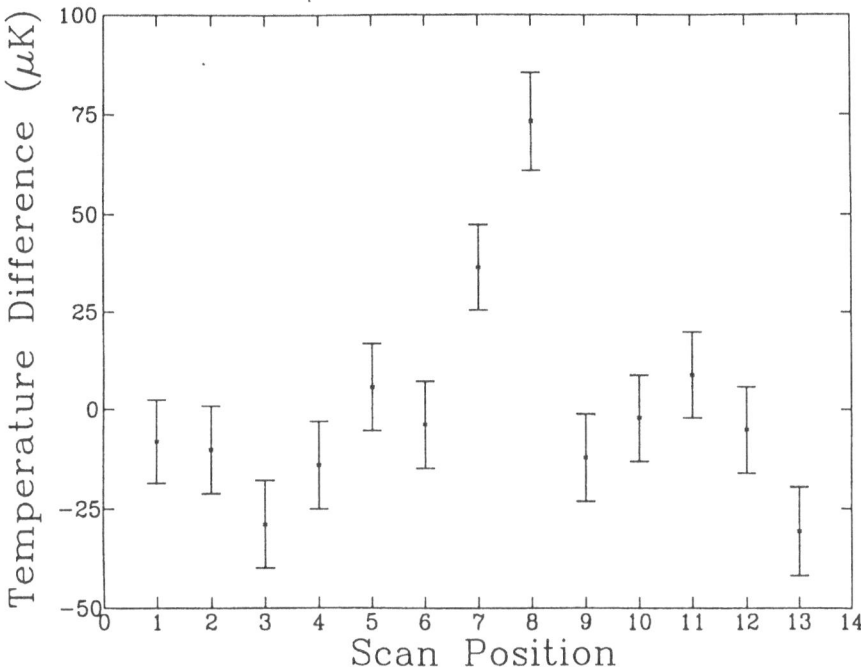

Fig. 5. Coadded data from Figure 4, under the assumption that all signal is due to CBR anisotropy

4 MAX 3

MAX consists of a multi-band dichroic bolometric photometer feeding the 1 meter Gregorian telescope described above. The system produces a FWHM of 0.5 degrees for each band. The elliptical secondary mirror nutates in a sinusoidal fashion to move the beam 1.3 degrees peak to peak on the sky. See Fischer et al (1992) and Meinhold et al (1993a) for more information about the MAX system. The data discussed below (the third flight of MAX) were taken with the system in a configuration using 3 bands, centered near 180, 270, and 360 GHz, with 20 to 30 % bandwidths. The photometer was cooled to 300 mK with a closed cycle ^{3}He refrigerator, and the detectors obtained a sensitivity to CBR fluctuations of 530, 770, and 2764 $\mu K\sqrt{sec}$ respectively.

The third flight of MAX took place on June 6, 1991 from Palestine, Texas. Several hours of high quality data were obtained, including two long integrations to search for fluctuations in the CBR. The data were deglitched to remove occasional cosmic ray and RF interference pulses, demodulated, and then coadded in azimuth angle on the sky. The measured statistical errors were comparable to the detector noise as measured in the laboratory. Calibrations of the system to antenna temperature were performed using a partially reflective membrane inserted into one lobe of the chopped beam at the focus. Jupiter was used to check this calibration. Data from this flight have been published in Devlin et al (1992), Meinhold et al (1993b), and Gundersen et al (1993).

4.1 Mu Pegasi

The longest integration was carried out around the star Mu Pegasi. The telescope was scanned smoothly $\pm 3°$ back and forth in azimuth for 1.4 hours. This integration occurred when Mu Pegasi was in the east, resulting in little rotation of the measured points relative to the telescope. Figure 6 shows the data from this integration, along with a calculation of the response of the MAX system to ISD emission derived from the high resolution IRAS 100 micron maps. The IRAS data have been scaled vertically for best fit for each band. There is an extremely good correlation between the IRAS data and the MAX measurements, leaving little doubt that most of the measured structure is due to ISD emission. The successful measurement of dust morphology at such a small level (10-50 μK) also demonstrates how far the system noise integrates down, and how well sidelobes and a number of other potential systematic problems have been controlled.

We have performed a variety of fits to the data of Figure 6 to ascertain what constraints they can place on CBR fluctuations. One method assumes two morphologically independent and spectrally distinct components contribute to the data. To constrain CBR fluctuations, we force the spectrum of one component to be that of a CBR anisotropy. These fits produce one component which is morphologically and spectrally like the ISD traced by the IRAS 100 μ maps, and a second component, possibly independent of the dust. This second component is not stable to the details of the data analysis, and is therefore only used as a conservative way of estimating upper limits to CBR anisotropy, by assuming that all the signal in this component is due to the CBR. This assumption leads to an upper limit to CBR fluctuations of $\Delta T/T < 2.35 \times 10^{-5}$ for a GACF at an angular scale of 25 arcminutes (the most sensitive scale). If the residual signal were assumed to be due to CBR anisotropy, it would imply $\Delta T/T = 1.4 \times 10^{-5}$ for a GACF at 25 arcminutes.

4.2 Gamma Ursae Minoris

The second long integration of the MAX 3 flight was carried out around the star Gamma Ursae Minoris (GUM). This region was chosen because a previous flight (Alsop et al, 1992) had shown evidence for structure with a spectrum consistent with CBR anisotropy and inconsistent with ISD emission. In addition, this is a region of sky with low dust emission, with dust contrast (as measured by the IRAS satellite at 100 μ) a factor of 5 below that in the region around Mu Pegasi. This region is located near the North celestial pole, and consequently rotates significantly relative to the local horizon during the measurement. A sample of 16 of the 39 pixels measured is shown in Figure 7. The essential features of the data set are evident in the figure. There is a very significant structure in the data which correlates well between channels, and decreases in amplitude with

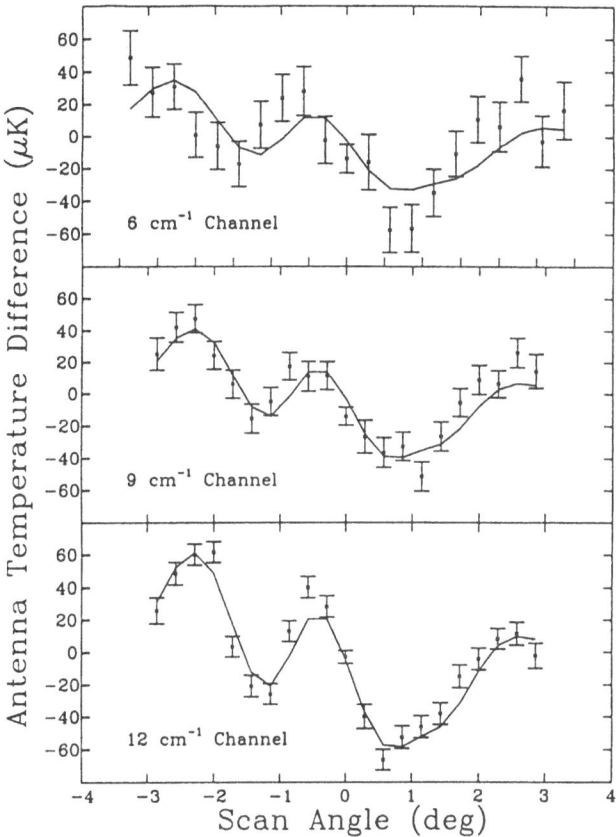

Fig. 6. MAX III data from measurement near the star Mu Pegasi (error bars). A model of dust emission based entirely on the IRAS high resolution 100 micron maps with a scale factor for each MAX channel is shown as solid lines. Note that the vertical axis is in Antenna temperature units: The 6,9, and 12 cm^{-1} channels need to be scaled by 2.3, 5.6, and 22.5 respectively to reference to a 2.7 K blackbody.

increasing frequency, as calibrated in antenna temperature. The spectrum of the fluctuations is well fit by a CBR spectrum. When coadded in CBR temperature units, the RMS of the observed structure is 145 μK, or $\Delta T/T = 5.3 \times 10^5$. Following is a brief discussion of possible origins of the structure other than CBR fluctuations.

Sidelobe contamination is considered an unlikely source of the structure. The best evidence for this is the Mu Pegasi data set discussed above. Those data were taken while the telescope tracked from 36° to 55° in elevation. Sidelobe contamination from the ground should be greater in the first half of the Mu Pegasi measurement than in the GUM measurement, while contamination from the balloon emission would be greater in the second half of the Mu Pegasi measurement than in the GUM measurement. The 95% confidence level upper limits to

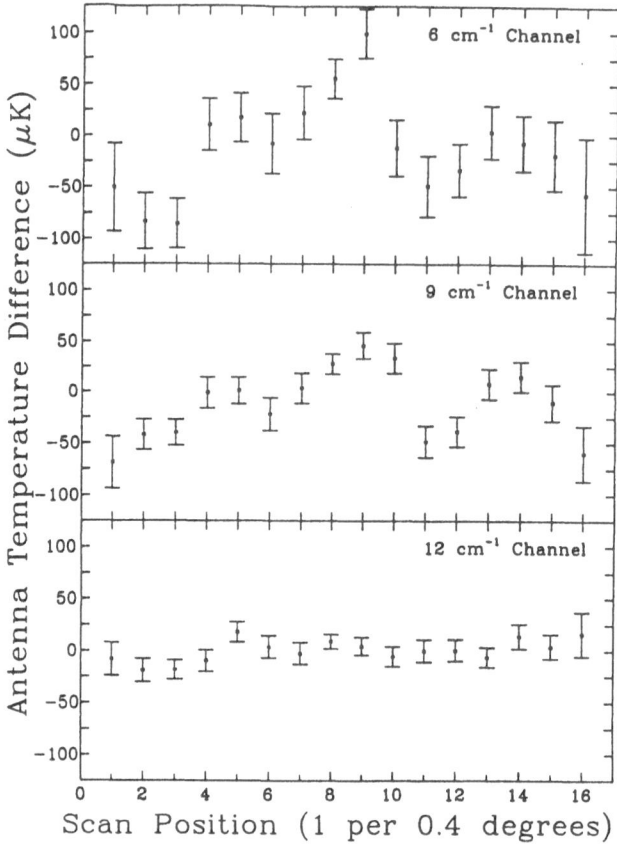

Fig. 7. 16 of 39 points from MAX III GUM data set. Vertical axis is again in antenna temperature units. Scaling to thermodynamic units is the same as Figure 6.

GACF CBR fluctuations for the Mu Pegasi data set first and second halves are 71 μK and 79 μK respectively, well below the measured signal.

The spectrum of the measured signal is extremely different from ISD. In addition, an extrapolation of the IRAS 100 μ data for GUM using the Mu Pegasi dust data predict very low differential dust emission.

Galactic synchrotron emission can be conservatively estimated by scaling the Haslam et al (1982) map, assuming antenna temperature scales as $\nu^{-2.7}$. This estimate provides less than 1% of the measured signal. Free-free emission is more difficult to estimate, but a similar conservative approach of assuming all signal in the Haslam map is due to free-free emission and scaling by $\nu^{-2.1}$ produces only 10% of the measured signal (and no significant morphological correlation).

Measurements of the CO(J=1-0) transition in this region (Wilson and Koch 1992, Thaddeus and Dame 1993) show there is no emission above 1 K km s^1. A 1 K km s^1 CO cloud filling a beam would give approximately a 10 μK signal at 6 cm^{-1} and a 5-10 μK signal at 9 cm^{-1}.

Figure 8 shows the weighted average of the 6 and 9 cm^{-1} data for the same set of points as Figure 7, rescaled to CBR thermodynamic temperature units. Under the assumption that the signal measured in the GUM region is due to CBR fluctuations, both upper and lower limits have been calculated for GACF at 25 arcminutes. At 95 % confidence level, the lower limit is $\Delta T/T > 3.1 \times 10^{-5}$, the most probable amplitude is $\Delta T/T = 4.2 \times 10^{-5}$ and the upper limit is $\Delta T/T < 5.9 \times 10^{-5}$.

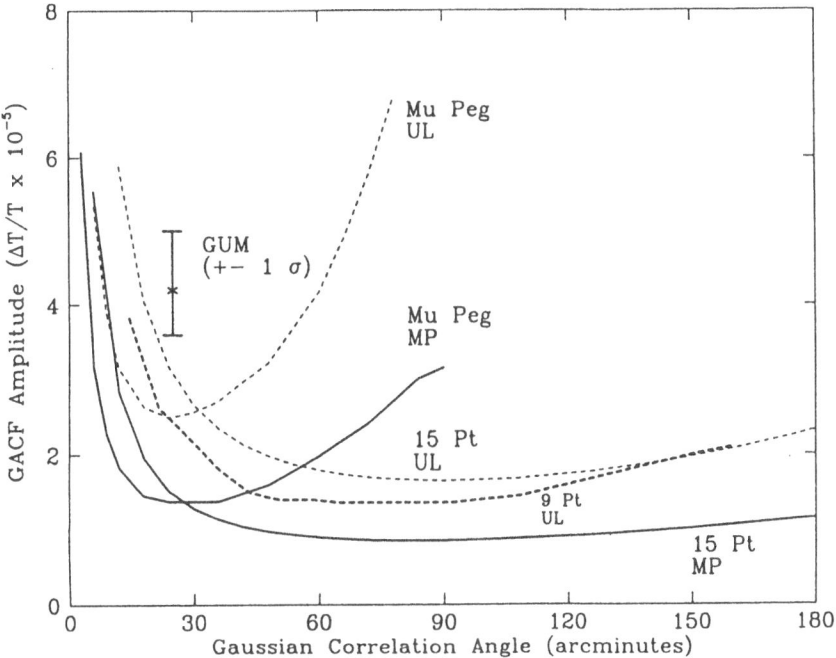

Fig. 8. 16 of 39 points from MAX III GUM data set, weighted average of 6 and 9 cm^{-1} channels. Vertical axis is in thermodynamic temperature units relative to a 2.73 K blackbody.

5 Comparisons and Discussion

Figure 9 shows all four data sets analyzed in the same way (using a GACF model). As noted earlier, the GACF is used primarily to allow a direct comparison among results at similar angular scales without reference to a cosmological model. It should be kept in mind that the upper limit curves are formally valid over the entire range, and exclude GACF skies with amplitudes larger than the curve at every point, at the 95% level. The detection or "most probable" curves are *not* valid over the entire range, but only pointwise. The best way to compare data sets is to compare the lowest point on the most probable curve to the same angular scale point on the upper limit curve. The most striking feature of the four data sets, aside from their sensitivity, is the apparent conflict between the two MAX 3 measurements, under the assumption of Gaussian distributed fluctuations described by a GACF. In fact, the upper limits from Gaier et al (1992) and Schuster et al (1993) are both also in apparent conflict with the GUM result. If the GUM data are correct, and are describing fluctuations in the CBR, then either our assumptions about the distribution of fluctuations are in question, or the sampling performed thus far in medium scale CBR anisotropy searches is inadequate. Note that we have only sampled less than 1% of the celestial sphere, with the combined data from both experiments.

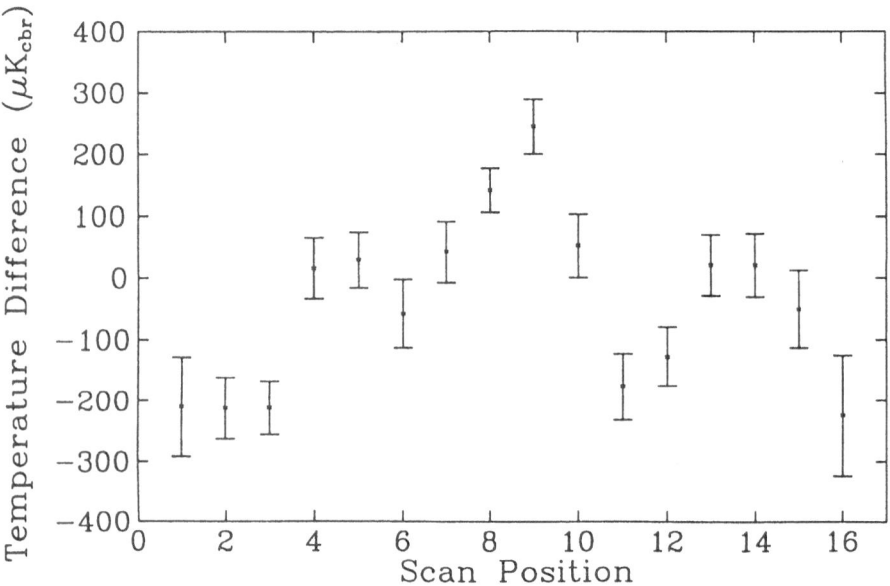

Fig. 9. Upper limits (UL) and most probable (MP) signals calculated from the data of Gaier et al (1992) (9 Pt), Schuster et al (1993)(15 Pt), Meinhold et al (1993) (Mu Peg) and Gundersen et al (1993)(GUM). The single point with error bars is the GUM detection, with $\pm 1\sigma$ errors. Dotted lines indicate 95% confidence level upper limits, solid lines indicate most probable signal levels (under the assumption that all relevant structure is due to CBR fluctuations).

The current generation of degree scale CBR anisotropy experiments has been forced to concentrate on extremely small areas of the sky, by a combination of foreground emission, available integration time (or detector sensitivity), and the need to control sources of systematic error extremely well. Currently planned array systems should help to alleviate the integration time and systematic error control problems at least.

6 Current Work

6.1 MAX 4

The MAX experiment flew in June, 1993 for the fourth time. In its latest incarnation, the detector was upgraded to an ADR, which kept the bolometers at a temperature of 85 mK, increasing the sensitivity. A fourth frequency channel was added near 90 GHz, to aid in discrimination of CBR from free-free emission. Three deep CBR integrations were performed, including one on the GUM region. Analysis of the data is proceeding.

This system is expected to fly again in May or June of 1994. Several more CBR anisotropy targets will be measured, as well as an attempt to measure the Sunyaev-Zeldovich effect towards Coma.

6.2 South Pole HEMT work

A new set of experiments is currently en route to the South Pole for an Austral summer season of observing. Two 1 meter Gregorian telescopes and a prime focus system will make observations with FWHM ranging from 1° to 4°. Multifrequency HEMT detectors similar to the one described above, but operating from 38-45 GHz and 26-36 GHz will be used on the telescopes.

6.3 Acknowledgements

The data reported here are the results of the concerted efforts of a large number of people over several years. Current contributors to the MAX effort include Andre Clapp, Mark Devlin, Marc Fischer, Andrew Lange, Paul Richards, and Stacy Tanaka of the Physics department of the University of California at Berkeley and Joshua Gundersen, Mark Lim, Philip Lubin, and P. Meinhold of the University of California at Santa Barbara.

The ACME-HEMT results are the work of Todd Gaier, Joshua Gundersen, Timothy Koch, Michael Seiffert, Jeffrey Schuster, Alexandre Wuenche, Philip Lubin and P. Meinhold, of the University of California at Santa Barbara.

All of those named above are also members of the NSF Center for Particle Astrophysics (CfPA).

This work has been supported by NASA under grant NAGW-1063; the NSF, under Polar grant NSF DPP 89-20578; and the Center for Particle Astrophysics. We wish to thank the entire ASA staff at the South Pole station for their excellent support during the 1990-91 season. We would also like to thank the NSBF crews for all the great flights, particularly June, 1991, and June, 1993. The K_a band HEMT amplifiers are on loan from the NRAO CDL. We are gratefull to Mike Balister, Marian Pospieszalski, and the staff of the NRAO CDL for their patience during numerous conversations on the subject of low noise amplification.

References

Alsop, D. C.,Cheng, E. S., Clapp, A. C., Cottingham, D. A., Fischer, M. L., Gundersen, J. O., Koch, T. C., Kreysa, E., Meinhold, P. R., Lange, A. E., Lubin, P. M., Richards, P. L., and Smoot, G. F. Astrophys. J. **317** (1992) 146

Devlin, M., Alsop, D., Clapp, A., Cottingham, D., Fischer, M., Gundersen, J., Holmes, W., Lange, A., Lubin, P., Meinhold, P., Richards, P., Smoot, G., 1992; Proc. Nat. Acad. Sci. USA (1992)

Fischer, M. L., Alsop, D. C.,Cheng, E. S., Clapp, A. C., Cottingham, D. A., Gundersen, J. O., Koch, T. C., Kreysa, E., Meinhold, P. R., Lange, A. E., Lubin, P. M., Richards, P. L., and Smoot, G. F. Astrophys. J. **388** (1992) 242

Gaier, T., Schuster, J. A., Gundersen, J. O., Koch, T., Seiffert, M. D., Meinhold, P. R., Lubin, P. M. Astrophys. J. **398** (1992) L1

Gundersen, J. O., Clapp, A.C., Devlin, M., Holmes, W., Fischer, M. L., Meinhold, P. R., Lange, A. E., Lubin, P. M., Richards, P. L., and Smoot, G. F.**413** L1-L5

Meinhold, P. R., and Lubin, P. M. Astrophys. J. **370** (1991) 11

Meinhold, P., Clapp, A., Devlin, M., Fischer, M., Gundersen, J., Holmes, W., Lange, A., Lubin, P., Richards, P., and Smoot, G. Astrophys. J. **409** (1993) L1

Meinhold, P. R., Chingcuanco, A. O., Gundersen, J. O., Schuster, J. A., Seiffert, M. D., Lubin, P. M., Morris, D., and Villela, T. Astrophys. J. **406** (1993) 12

Pospieszalski, M. W., Gallego, J. D., Lakatosh, W. J. Proc. 1990 MTT-S Int. Microwave Symp. (1990) 1253.

Schuster, J., Gaier, T., Gundersen, J., Meinhold, P., Koch, T., Seiffert, M., Wuensche, C., and Lubin, P. Astrophys. J. **412** (1993) L47

Results from the Cosmic Background Explorer[1]

G.F.Smoot[2]

[1] The National Aeronautics and Space Administration/Goddard Space Flight Center (NASA/GSFC) is responsible for the design, development, and operation of the Cosmic Background Explorer (COBE). Scientific guidance is provided by the COBE Science Working Group.
[2] Lawrence Berkeley Laboratory, Berkeley, CA 94720, USA

ABSTRACT - *COBE* has produced significant new scientific findings and results in the past year. The DMR instrument reported detections of temperature fluctuations in the cosmic microwave background (CMB) radiation. The DMR data are consistent with power law spectrum of Gaussian initial fluctuations with the quadrupole-normalized amplitude of $Q_{rms-PS} = 17\pm3$ μK and a power law index $n = 1.1 \pm 0.6$. These data are supportive of models of structure formation through gravitational instability.

The FIRAS instrument results improved indicating that the spectrum of the cosmic microwave background deviates from a Planckian spectrum by less than one part in 3000 of the peak intensity over the wavelength range 0.05 to 5 mm. These data rule out a number of alternative models of structure formation including explosive scenarios.

The initial *COBE* data products, in particular the first year DMR sky maps, were released to the scientific community in June 1993.

1 Introduction

The origin of large scale structure in the Universe is one of the most important issues in cosmology. Currently the leading models for structure formation postulate gravitational instability operating upon a primordial power spectrum of density fluctuations. The inflationary model of the early Universe [21] produces primordial density fluctuations [2],[22] [24],[39] with a nearly scale-invariant spectrum suggesting a viable mechanism for structure formation. Structure forms as the result of gravitational amplification of initially small perturbations in the primordial mass-energy distribution, and non-baryonic dark matter seems necessary to provide sufficient growth of these perturbations. The determination of the nature of the initial density fluctuations then becomes an important constraint to cosmological models [8]. The discovery of the anisotropy in the cosmic microwave background radiation by the COBE DMR instrument [36],[3],[25], [45] and the recent confirmation [18] mark a new era in cosmology and the beginning of investigations of these primordial fluctuations.

2 DMR and CMB Anisotropy

The Differential Microwave Radiometer (DMR) experiment is designed to map the microwave sky and find fluctuations of cosmological origin. For the 7° angular scales observed by the DMR, structure is superhorizon size so the spectral and statistical features of the primordial perturbations are preserved [32]. The DMR maps the sky at frequencies of 31.5, 53, and 90 GHz (wavelengths of 9, 5.7, and 3.3 mm). The frequency independence of the anisotropy is a strong argument that the anisotropy is in the cosmic microwave background radiation and not due to foreground Galactic emission or extragalactic sources. The confirming 'MIT' balloon-borne bolometer observations [18] have an effective frequency of about 170 GHz making the argument stronger. The typical fluctuation amplitude is roughly 30 μK or $\Delta T/T \sim 10^{-5}$ on a scale of 10°. The data appear consistent with a scale-invariant power spectrum with an uncertainty of ± 0.6 in the exponent of the power law. The amplitude and spectrum are consistent with that expected for gravitational instability models involving nonbaryonic dark matter, and perhaps consistent with the measured large scale velocity flows. The angular power spectrum of the DMR maps is estimated by several methods, and is consistent with a near scale invariant power spectrum drawn from a Gaussian distribution. The horizon due to expansion of the universe limits the region over which we can observe these primordial fluctuations. For the very largest scales only a small number of fluctuations will be present inside our horizon creating error due to our cosmic sampling variance. If the fluctuations are drawn from a Gaussian distribution, the cosmic variance will limit the DMR's ability to determine the mean cosmic fluctuation amplitude to about 10%.

Since the DMR detection was announced several medium and smaller angular scale experiments have new results. These include: the UCSB South Pole experiment [17],[35], the Princeton Saskatoon experiment [44], the Tenerife collaboration [27],[42], the CARA South Pole experiments: Python [13], and 'White Dish' [33], the Center for Particle Astrophysics MAX balloon-borne experiment [12],[19],[30], the MSAM balloon-borne experiment [9], the Owens Valley Radio-astronomy Observatory (OVRO) [31], the Roma balloon-borne experiment ULISSE [11], and the Australia Telescope [41]. In general, these experiments report fluctuations at the 10^{-5} level. These measurements could in principle distinguish among models of structure formation, shed information on the nature of the dark matter, and probe the existence of cosmological gravity waves [10], [37]. However, the current results vary at the factor of two to three level which is just what is needed to make the distinctions. There is perhaps evidence that the data are not self-consistent. In part one can assume that some discrepancy is due to experimental error, to the limited region of the sky sampled by these experiments and to the potential confusion by galactic emission and extragalactic point sources. The community eagerly awaits refined observations of the power spectrum. In order to distinguish the various model one needs to measure the large angular scale power spectrum as accurately as possible given cosmic variance and utilize that as a normalization to the primordial spectrum against which to compare other observations and theoretical models.

3 CMB Fluctuation Power Spectrum

The initial results of the DMR data analysis [36] showed that the two-point correlation function was well fitted by a power law power spectrum with the quadrupole-normalized amplitude of $Q_{rms-PS} = 17\pm3$ μK and a power law index $n = 1.1 \pm 0.6$. We have continued and extended the analysis taking more care in the analysis in anticipation of additional years' data and utilizing the existing data more fully. There are several corrections and foregrounds that must considered if one is to determine things to the 10% level.

One item that must be treated more carefully is the DMR beam filter function. For the first cut analyses we approximated the DMR antenna beam pattern as a Gaussian. An undergraduate, Ruediger Kneissl and I have done a more detailed treatment of the actual pattern and the resulting filter function. The correct filter function depends upon the particular data set being analyzed: antenna beam only, including 0.5-sec integration, including 2.6° pixel size, and 2.6° correlation function bins. The filter function for the beam is publised [47] and the others are available on request. The net effect of using the actual filter rather than the Gaussian approximation is to boost the intermediate, $(10 < l < 30)$, angular scale power and thus raise the spectral index by 0.2 to 0.3.

At a smaller level we are concerned with the effects of galactic cuts and uneven sky coverage. In this volume Tenorio et al. discuss the effect of these upon

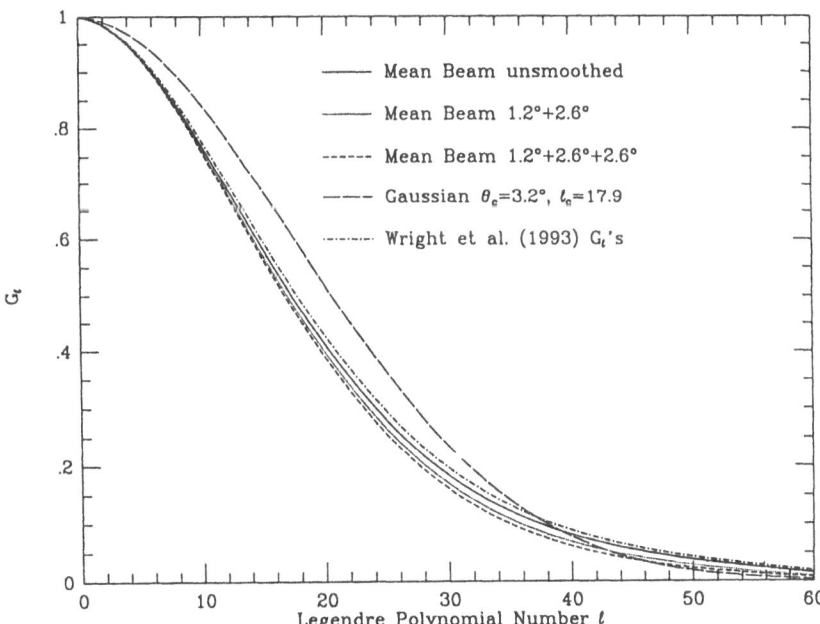

Fig. 1. Antenna Filter Function for various levels of data processing. These include the beam, the beam with 0.5-second integration time, plus 2.6° pixelization, and plus 2.6° correlation function binning. These curves from memo by Kneissl and Smoot.

the quadrupole amplitude. The effect on the full power spectrum is significantly less but still must be considered at the per cent level.

Because the DMR measures differences of all the pixels that are 60° apart on the sky, the instrument noise results in a slight correlation in the data at a 60° separation at the few per cent level. Because of cuts and uneven sky coverage this effect is mixed to other angles at the per cent or lower level. This angular correlation of noise shows up in the autocorrelation function but not in the cross-correlation of channels or maps with independent noise. These questions of the correlation of pixels noise and errors due to the differencing and map making procedure and the cuts,uneven sky coverage, and weighting are a major part of the thesis work of my graduate student, Charles Lineweaver. The broad strokes of these effects have been explored and we are at the tidying and writing stage.

Another major area of concern is the question of galactic and extragalactic foregrounds. In the discovery papers [36],[3], [25],[45] we estimated that these galactic and extragalactic emissions were at the or less than 10% of the signal we observed. We have continued to work and gathered new data on the galactic emission. It appears that the galactic emission may be less anisotropic than feared and that the free-free emission in particular may be smoother and less intense than allowed. On the DMR's large angular scale extragalactic point-like sources are not significant. The cosmic nature of the measured fluctuations was tested [4] by correlating the DMR maps with maps of Galactic emission, the X-ray background, Abell clusters, and other foregrounds. No evidence for significant correlation was found. The "blamb" structure reported by the Relikt team [40] is not present in the DMR map. The net result is that except for the quadrupole the galactic emission effects are likely to be smaller than 10%.

The remaining question is what are the residual systematic errors? Have we found any new ones? And finally is there some left undiscovered? We have continued our data processing and analysis and have a new set of data processing software and have recently completed running the first two years of DMR data. We are now checking and verifying that the runs were done correctly and then will move into investigating the systematic errors. Our early and preliminary analysis indicate that the residual systematic errors are still not significant.

Our continuing analysis show the data are still consistent with a power law spectrum with previously quoted parameters: the quadrupole-normalized amplitude of $Q_{rms-PS} = 17\pm3$ μK and a power law index $n = 1.1 \pm 0.6$. We have used a number of approaches to determine the power spectrum directly including fitting spherical harmonics directly to the map, fitting the two-point correlation function to Legendre polynomials, and fitting the power spectrum to the map directly. We have also used topological measures and the sky-rms as a check of the statistics and mean power spectrum in the map. The results are all generally consistent. The DMR first year data power spectrum is shown along with the data from other recent CMB anisotropy experiments in Figure 2.

We have reason to hope that with four years of DMR data that the cosmic variance will be the dominant uncertainty in determining the CMB fluctuation power spectrum.

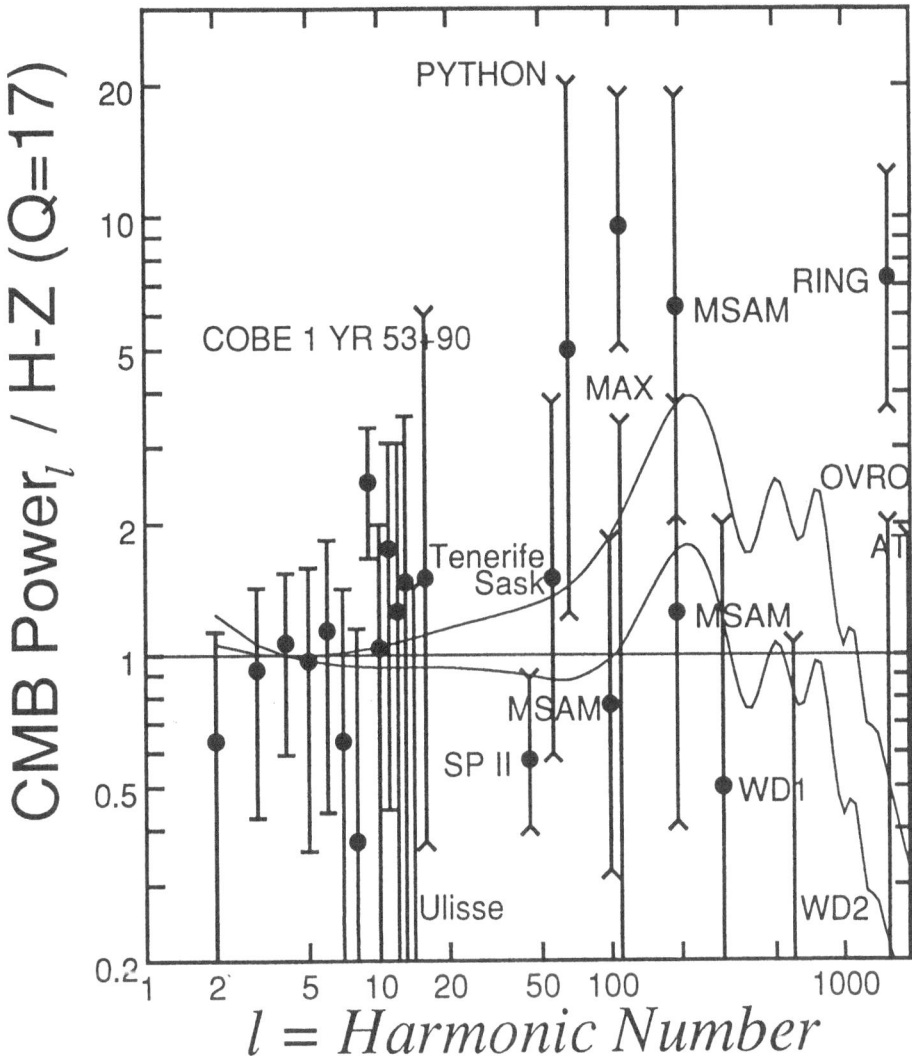

Fig. 2. Power Spectrum from COBE DMR and other Anisotropy Experiments. The data points refer to the experiments mentioned in ssection 1 and the two curves are from Crittendon et al. 1993 representing the predicted spectrum for inflation generated (upper) pure density fluctuations (essentially a flat inflaton potential) and (lower) equal density and gravitational wave contributions to the quadrupole.

4 FIRAS and CMB Spectrum Observations

The Far Infrared Absolute Spectrophotometer (FIRAS) instrument compared the spectrum of the CMBR to that of a precise blackbody for the first time. It has a $7°$ diameter beamwidth, and covers two frequency ranges, a low frequency channel from 1 to 20 cm^{-1} and a high frequency channel from 20 to 100 cm^{-1}. Preliminary results [28] showed that the CMBR is consistent with a blackbody at 2.735 ± 0.06 K, and that deviations are less than 1% of the peak brightness. The UBC rocket result [20] nearly immediately confirmed the FIRAS results. New FIRAS results show that deviations from a blackbody are 30 times smaller: less than one part in 3000 of the peak intensity [14],[15],[28],[47].

The absolute temperature of the cosmic background, T_0, was determined in two ways. The first uses the thermometers in the external calibrator and gives $T_0 = 2.730$ K. The second calibrates the temperature scale from the wavelength scale, and gives 2.722 K for T_0. The adopted value is 2.726 ± 0.010 K (95% confidence [29]), which averages these two methods. Three additional determinations of T_0 depend on the dipole anisotropy. The spectrum of the dipole anisotropy is sensitive to the assumed blackbody temperature. Since the velocity of the solar system with respect to the CMB is not known a priori, only the shape and not the amplitude of the dipole spectrum can be used. For FIRAS, this analysis [14] gives $T_0 = 2.714 \pm 0.022$ K, while for DMR [26] it gives $T_0 = 2.76 \pm 0.18$ K. The DMR data analysis keeps track of the changes in the dipole caused by the variation of the Earth's velocity around the Sun during the year. In this case the velocity is known, so T_0 can be determined from the amplitude of the change in the dipole, giving $T_0 = 2.75 \pm 0.05$ K.

The spectrum observations imply a tight limit on energy release in the early universe and strong support for the hot Big Bang model. From a redshift of about 10^6 to 10^3 no process can release electromagnetic energy at a level exceeding about 10^{-4} of that in the cosmic background radiation. Limits on the distortion parameters are $|y| < 2.5 \times 10^{-5}$ and $|\mu/kT| < 3.3 \times 10^{-4}$ with 95% confidence. The Comptonization parameter y restricts the possible thermal history of the intergalactic medium, which must not be very dense or very hot (less than $\approx 10^4$ KeV). In addition, the FIRAS results limit energy release into the far infrared from Population III stars or evolving IRAS galaxies. In both cases, less than 1% of the hydrogen could have burned [47] after a redshift of 80, assuming $\Omega_{baryon} h^2 = 0.015$.

The FIRAS results can be combined and compared with other observations of the cosmic background spectrum. At this time the spectrum of the cosmic background is well described by a single temperature blackbody over four decades in frequency (or wavelength), without significant deviations. However, more precise measurements at long wavelengths could improve the COBE limits on μ by a factor of 10.

Non-cosmological results of the FIRAS include the determination of the mean far infrared spectrum of the Galaxy, and its decomposition into two components of dust emission and 9 spectrum lines [45]. The lines of [N II] and [C II] have been further interpreted [5], [34].

5 DIRBE and the Cosmic Infrared Background

The primary objective of the Diffuse Infrared Background Experiment (DIRBE) is to conduct a definitive search for an isotropic cosmic infrared background (CIB), within the constraints imposed by the local astrophysical foregrounds, from 1 to 240 μm . Additional objectives include studies of the interplanetary dust cloud and the stellar and interstellar components of the Galaxy. Both the cosmic redshift and the reprocessing of short-wavelength radiation to longer wavelengths by dust act to shift the short-wavelength emissions of cosmic sources toward or into the infrared, and the CIB may contain much of the energy released since the formation of luminous objects. Measurement of the CIB would provide important new insights into issues such as the amount of matter undergoing luminous episodes in the pregalactic Universe, the nature and evolution of such luminosity sources, the nature and distribution of cosmic dust, and the density and luminosity evolution of infrared-bright galaxies.

The DIRBE has obtained absolute brightness maps of the full sky in 10 photometric bands (1.2, 2.2, 3.5, 4.9, 12, 25, 60, 100, 140 and 240 μm). To facilitate discrimination and study of the bright foreground contribution from interplanetary dust, linear polarization is measured at 1.2, 2.2, and 3.5 μm , using a combination of orthogonal polarizers and spacecraft rotation. All celestial directions are observed hundreds of times at all accessible angles from the Sun in the range 64 - 124°. The instrument has a large field of view, 0.7°square, and the sky signal is continuously chopped against a zero-flux internal surface. A cold shutter allows measurement of instrumental offsets and internal stimulation of the detectors.

The photometric quality of the DIRBE data is excellent; when the full reduction of the cryogenic-era data is complete, photometric consistency over the sky and over the 10 month period is expected to be near 1% or better. The instrument rms sensitivity per field of view in 10 months is $\lambda I_\lambda = (1.0, 0.9, 0.6, 0.5, 0.3, 0.4, 0.4, 0.1, 11.0, 4.0) \times 10^{-9}$ W m^{-2} sr^{-1}, respectively for the ten wavelength bands listed above. These levels are generally well below estimated CIB radiation contributions and foregrounds.

Papers on the foregrounds have been submitted to the Astrophysical Journal and presented [1],[6],[16], [23],[38],[43] at the Back to the Galaxy Conference. Preliminary full sky maps at wavelengths from 1.2 to 240 μm have provided dramatic new views of the stellar and interstellar components of the Milky Way. The zodiacal dust bands discovered in the IRAS data are confirmed, and scattered near-infrared light from the same particles has also been detected. Starlight from the galactic bulge region, after correction for extinction, has been shown to have an asymmetric distribution consistent with a non-tilted stellar bar. The warp of the near and far infrared emission near the galactic plane is similar to that expected from previous studies of the stellar and interstellar components of the Galaxy. New upper limits have been set on the CIB all across the infrared spectrum, conservatively based upon the minimum observed sky brightness [23].

6 COBE Data Products Release

An initial set of COBE data products from all three instruments was released in June 1993, and a new data release in June 1994 will include all-sky DIRBE and FIRAS coverage, DIRBE polarimetry, FIRAS data from the low-frequency band, and the first two years' worth of DMR data. Additional data will be released in June 1995. Documentation and initial data products are available by anonymous FTP from nssdca.gsfc.nasa.gov with the username "anonymous" and your e-mail address as password. Change to directory [000000.cobe] and get the file aareadme.doc. Data and documentation may also be obtained on tape by request to the Coordinated Request and User Support Office (CRUSO), NASA/GSFC, Code 633.4, Greenbelt, MD 20771, phone: 301-286-6695, e-mail: request@nssdca.gsfc.nasa.gov.

7 Discussion and Summary

The COBE has been a remarkably successful space experiment with dramatic observational consequences for cosmology, and the DIRBE determination of the cosmic infrared background is yet to come. The very tight limits on deviations of the spectrum from a blackbody rule out many non-gravitational models for structure formation, while the amplitude of the ΔT discovered by the COBE DMR implies a magnitude of gravitational forces in the Universe sufficient to produce the observed clustering of galaxies, but perhaps only if the Universe is dominated by dark matter. The DMR ΔT provides measurement of the 'initial conditions' for the gravitational instability modes.

ACKNOWLEDGEMENTS - COBE was supported by the NASA Office of Space Sciences. COBE was a large team effort involving many engineers, scientists, analysts and others. I would like to thank Enrique Martinez-Gonzalez, Laura Cayon, and Jose Luis Sanz for arranging and managing the Santander Workshop and acknowledge their and Luis Tenorio's efforts in getting out this proceedings. This work supported in part at LBL through DOE Contract DOE-AC-03-76SF0098.

References

1. Arendt, R.G., et al., Back to the Galaxy, eds. S.S. Holt and F. Verter, (New York: AIP Conf. Proc.), (1993).
2. Bardeen, J. M., Steinhardt, P. J. & Turner, M. S. 1983, Phys. Rev. D, 28, 679
3. Bennett, C. L., et al. 1992, ApJ, 396, L7.
4. Bennett, C. L., et al., 1993, ApJ, COBE Preprint 93-08.
5. Bennett, C.L., and Hinshaw, G., Back to the Galaxy, eds. S.S. Holt and F. Verter, (New York: AIP Conf. Proc.), (1993).
6. Berriman, G.B., et al., Back to the Galaxy, eds. S.S. Holt and F. Verter, (1993).
7. Boggess, N. et al., 1992, ApJ, 397, 420.
8. Bond, J. R., & Efstathiou, G. 1987, MNRAS, 226, 655

9. Cheng, E. S., Cottingham, D. A., Fixsen, D. J., Inman, C. A., Kowitt, M. S., Meyer, S. S., Page, L. A., Puchalla, J. L. & Silverberg, R. F. 1993 preprint

10. Crittenden, R., Bond, J. R., Davis, R. L., Efstathiou, G. & Steinhardt, P. J. 1993, PRL, 71, 324-327.

11. de Bernardis, P., Masi, S., Melchiorri, F., Melchiorri, B. & Vittorio, N. 1992, ApJ, 396, L57-L60.

12. Devlin et al 1993, preprint.

13. Dragovan, M. et al. 1993, private communication.

14. Fixsen, D.J., et al. 1993a, COBE Preprint 93-04, ApJ, Jan 10, 1994.

15. Fixsen, D.J., et al. 1993b, COBE Preprint 93-02, ApJ, Jan 10, 1994.

16. Freudenreich, H.T., et al., Back to the Galaxy, eds. S.S. Holt and F. Verter, (1993).

17. Gaier, T. et al, Ap.J., 398, L1 (1992).

18. Ganga, K. et al, 1993, Ap.J.,410, L57

19. Gunderson et al, 1993, Ap.J.,413, L1

20. Gush, H. P., Halpern, M., and Wishnow, E. H., 1990, PRL, 65, 537.

21. Guth, A. 1981, Phys. Rev. D, 23, 347.

22. Guth, A. & Pi, Y-S., 1982, PRL, 49, 1110

23. Hauser, M.G. Back to the Galaxy, eds. S.S. Holt and F. Verter, (1993).

24. Hawking, S., 1982, Phys. Lett., 115B, 295

25. Kogut, A., et al. 1992, ApJ, 401, 1.

26. Kogut, A., et al. 1993, ApJ, 419, Dec 10,

27. Hancock, S., et al. 1993, preprint, submitted to Nature.

28. Mather, J.C. et al. 1990, ApJL, 354, L37-L41.

29. Mather, J.C., et al. 1993, COBE Preprint 93-01, ApJ, Jan 10, 1994.

30. Meinhold, P., Clapp, A., Devlin, M., Fischer, M., Gundersen, J., Holmes, W., Lange, A., Lubin, P., Richards, P. & Smoot, G. 1993, ApJL, 409, L1-L4.

31. Myers, S. T., Readhead, A. C. S., & Lawrence, C. R. 1993, ApJ, 405, 8-29.

32. Peebles, P.J.E., 1980, Large Scale Structure of the Universe, Princeton Univ. Press, 152

33. Peterson, J. et al. 1993, private communication.

34. Petuchowski, S. and Bennett, C.L., 1993, ApJ, 405, 591.

35. Schuster, J., Gaier, T., Gundersen, J., Meinhold, P., Koch, T., Seiffert, M., Wuensche, C. & Lubin P. 1993, ApJ, 412, L47-L50.

36. Smoot, G. F., et al. 1992, ApJ, 396, L1

37. Smoot, G. F., and Steinhardt, P. 1993, J. Quantum and Classical Gravity.

38. Sodroski, T.J., et al., Back to the Galaxy, eds. S.S. Holt and F. Verter, (1993).

39. Starobinskii, A.A., 1982, Phys. Lett., 117B, 175

40. Strukov, I.A, et al, 1992, MNRAS, 258, 37P.

41. Subrahmayan, R., Ekers, R. D., Sinclair, M. & Silk, J. 1993, MNRAS, 263, 416-424.

42. Watson, R. A., Gutierrez de la Cruz, C. M., Davies, R. D., Lasenby, A. N., Rebolo, R., Beckman, J. E. & Hancock, S. 1992, Nature, 357, 660-665.

43. Weiland, J.L., et al., Back to the Galaxy, eds. S.S. Holt and F. Verter, (New York: AIP Conf. Proc.), (1993).

44. Wollack, E. J., Jarosik, N. C., Netterfield, C. B., Page, L. A. & Wilkinson, D. T. 1994, ApJL.

45. Wright, E.L., et al., 1991, ApJ, 381, 200.

46. Wright, E.L., et al., 1992, ApJ, 396, L13

47. Wright, E.L. et al. 1993, COBE preprint 93-03, ApJ, Jan 10, 1994.

48. Wright, E.L. et al. 1993, COBE preprint 93-06, ApJ.

The MSAM/TopHat Program for Measuring the CMBR Anisotropy *

Edward S. Cheng

NASA/Goddard Space Flight Center, Code 685, Greenbelt MD 20771, USA

1 Abstract

We describe a series of three complementary balloon-borne measurements of the anisotropy in the Cosmic Microwave Background Radiation (CMBR) and astrophysical foreground sources on angular scales from 0.5° to 180° at frequencies between 70 and 700 GHz. On the largest angular scales, we plan to finish mapping the entire sky in the far-infrared with 3.8° resolution and sensitivity comparable to the *COBE*/DMR first year maps. The analysis of the first flight of this Far-Infrared Survey (FIRS) led to the confirmation of the *COBE* anisotropy discovery. On the 0.5° to 3° angular scale we will refine our pointing telescope and make an additional flight of the Medium-Scale Anisotropy Measurement (MSAM) using the existing radiometer. This will strengthen the results obtained from the first flight (Cheng *et al.* 1993a). Two subsequent flights, observing the same sky positions with a lower frequency multi-channel radiometer (MSAM II), will ensure that the result is not contaminated by Galactic or extragalactic foreground radiation. Concurrent with the above program, we will implement a new concept in CMBR measurements which will provide the next step in observational capabilities. This experiment, named TopHat, will be mounted on top of a balloon and will observe for two weeks in a circumpolar flight launched from Mc-Murdo Station, Antarctica. By providing a combination of reduced systematics and extended integration time, it offers a factor of 30 improvement in sensitivity to CMBR anisotropy over existing measurements.

2 Introduction

The recent discovery of the anisotropy in the Cosmic Microwave Background Radiation (CMBR) by the *COBE* satellite (Smoot *et al.* 1992, Bennett *et al.* 1992) has begun a new era in quantifying the density fluctuations which gave rise to

---

* This research program is carried out by a collaboration with members from NASA/GSFC, Bartol Research Institute, Brown University, University of Chicago, and Princeton University.

structure in the Universe. The experience gained from this effort, coupled with major advances in detector sensitivity, improved measurements of astrophysical foreground radiation, and a more complete understanding of systematic and instrumental effects, places us on the threshold of revealing the detailed spectrum of these fluctuations. Such measurements are needed to provide the fundamental observational constraints for building models of large-scale structure formation. Our program concentrates on the large ($3°$ to $180°$) and medium ($\sim 1°$) anisotropy scales, which probe the primordial fluctuations and the modifications of these fluctuations at the surface of last scattering, respectively. The comparison of results from these two regimes will provide constraints for basic cosmological constants, especially Ω_{baryon} (Gorski 1993) and the shape of the primordial density fluctuation spectrum (Kashlinsky 1992, Efstathiou et al., 1992).

Sensitivity improvements in CMBR anisotropy measurements have been a remarkable success story. The past decade has brought over a factor of ten increase in sensitivity, enabled by the development of new technologies and, in particular, the successful application of low-noise bolometric detectors (Page et al. 1990, Meyer et al. 1991, Ganga et al. 1993a, Ganga et al. 1993b, Readhead et al. 1989, Meinhold & Lubin 1991, Gaier et al. 1992, Fischer et al. 1992, Myers et al. 1992, Watson et al. 1991, Meinhold et al. 1993). In spite of these advances, the first widely-accepted detection of anisotropy in the CMBR was made with the *COBE/DMR* instrument, which is nearly a factor of ten less sensitive than those in common use today. While experiments with greater sensitivity have detected significant structure on both large and intermediate angular scales (Meyer et al. 1991, Meinhold et al. 1993, Gundersen et al. 1993, Cheng et al. 1993a), an unequivocal detection of CMBR anisotropy must be based on a careful understanding of systematic errors and possible foreground sources (Bennett et al. 1992, Kogut et al. 1992). It is therefore imperative that all planned measurements of CMBR anisotropy address these concerns directly, even at the expense of raw detector sensitivity so long as the overall *system* sensitivity is sufficient to provide new insights. We emphasize the need for improved *system* sensitivities, which include all uncertainties from modeling unknown parameters of astrophysical foreground sources (Brandt et al. 1993), as well as possible systematic effects of the measurement strategy.

Our program, summarized here, achieves a set of complementary scientific objectives with three balloon-borne measurements of the CMBR anisotropy. The details for each measurement are provided in subsequent sections.

1. Far-Infrared Survey (FIRS). This is a large-scale survey of the far-infrared sky in four bands between 5 and 23 cm^{-1} (150 and 690 GHz) with an angular resolution of $3.8°$. We plan to fly the existing package three more times, once in the Northern hemisphere, and twice in the Southern hemisphere. Together with the existing Northern hemisphere data, this will produce an all-sky map with sensitivity comparable to the *COBE*/DMR first-year maps but at a factor of two finer resolution and a factor of three higher frequency. This experiment has confirmed the DMR detection using data from the first Northern hemisphere flight.

2. Medium-Scale Anisotropy Measurement (MSAM). The first phase of this project (MSAM I) uses the same radiometer as the FIRS (and thus the same spectral bands). Combined with a pointed telescope it measures anisotropy on 0.5° to 3° angular scales. The first flight of this package has resulted in a detection at the $\delta T/T \sim 2 \times 10^{-5}$ level. MSAM I will be completed after one additional flight. Subsequently, two flights are planned with the same gondola, but using a lower frequency radiometer (MSAM II). Working between 70 and 150 GHz, a powerful observational check for foreground sources will be made by observing the same sky positions as MSAM I while simultaneously improving sensitivity to CMBR anisotropy by at least a factor of three.

3. Long Duration Balloon Measurement (TopHat). Designed to characterize (not just detect) CMBR anisotropy, it is optimized to reject spurious detections, both systematic and foreground. This is achieved by placing the telescope above the balloon, which provides a unique observing environment unequaled in any other near-Earth platform. A test flight of TopHat will occur after a three year development effort. Shortly following this test flight, we will launch TopHat from McMurdo Station, Antarctica, for a two week circumpolar flight. Observing in five bands between 5 and 21 cm^{-1} (150 and 630 GHz), this system will measure 40 points on the sky, each with an RMS δT of 1 μK ($\delta T/T \sim 3 \times 10^{-7}$) after foreground modeling. By operating in a long duration ballooning environment on the top of a balloon, an exceptional opportunity is created which allows for systematic error checking in a configuration that minimizes the problems which have plagued previous balloon measurements.

3 Spectral Coverage

Measurements near the peak of the CMBR anisotropy spectrum provide a unique way of distinguishing between true CMBR anisotropy and foreground emission. Very few foreground sources have the characteristic rollover of the CMBR anisotropy spectrum, tending instead to be monotonic in the spectral region near the peak of the CMBR (Figure 1). The ability to discriminate against foregrounds is an essential feature of all well-designed CMBR anisotropy measurements and has motivated our efforts to design optimized experiments which include measurements around the CMBR peak at 180 GHz (6 cm^{-1}). While measurements above 90 GHz preclude operating in the experimentally less challenging ground-based mode because of atmospheric noise, the unique spectral signature of the CMBR anisotropy peak makes this region especially attractive.

The dominant contributions to the sky brightness at millimeter and submillimeter wavelengths are the CMBR itself, and Galactic emission in the form of synchrotron radiation, bremsstrahlung, and thermal dust emission. Extragalactic radio sources are also potentially troublesome. Figure 1 shows the spectrum for the Galactic foreground sources assuming a 10% contrast in the average (high-Galactic latitude) brightness difference between nearby ($\sim 1°$) patches in the sky.

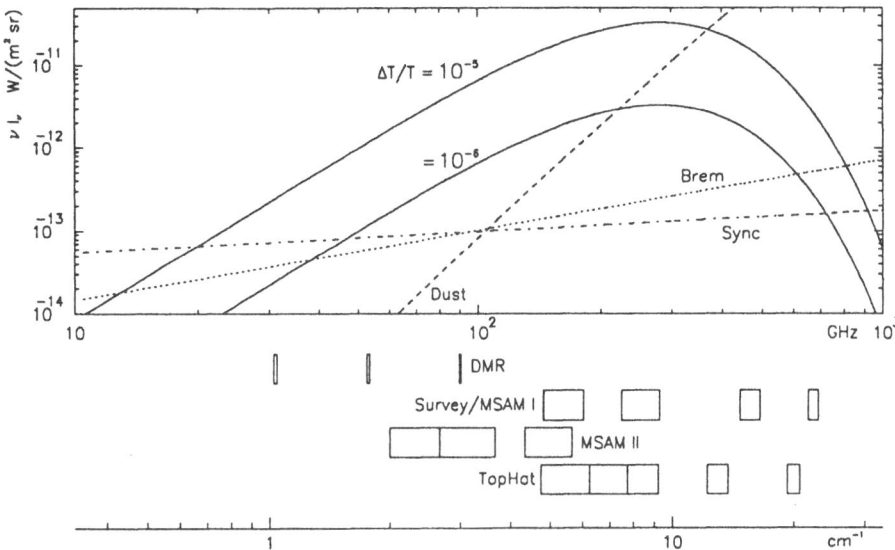

Fig. 1. Differential Spectra of Foreground Sources. The two solid curves show the CMBR anisotropy spectrum for $\delta T/T = 10^{-5}$ and 10^{-6}. The dashed, dot, and dot-dash curves are dust emission, bremsstrahlung, and synchrotron radiation for high Galactic latitude, based on Bennett *et al.* 1992. A 10% contrast in the foreground sources has been assumed. Also shown are the channel placements and bandwidths for the FIRS/MSAM I, *COBE*/DMR instruments, and the MSAM II and TopHat radiometers.

When extracting the cosmological signal, a sky model must be used to decompose the measured spectrum into CMBR and foreground components. Clearly, the number of model components that can be determined is bounded by the number of spectral bands. The placement of these bands can and must be optimized for the best discrimination of the CMBR from dominant foregrounds. From Figure 1, the dust emission is the most prominent foreground source at frequencies near and above the CMBR peak, but its spectrum is sufficiently different that it can be well separated from the CMBR using spectral bands at $\nu \gtrsim 10$ cm^{-1}. This is true even if the dust must be described by a two-component model which includes a cold component (Wright *et al.* 1991). Regions in the sky with substantial dust emission can be identified from the *COBE*/DIRBE and IRAS 100 μm surveys and avoided for CMBR studies.

Synchrotron radiation is expected to dominate the foregrounds only at frequencies below 80 GHz, and should exceed $\delta T/T \sim 10^{-6}$ CMBR anisotropy only below 50 GHz. The synchrotron radiation model of Bennett *et al.* 1992, based on radio surveys at 408 MHz and 1420 MHz together with the local electron spectrum, provides a reasonable estimate for the contamination due to this foreground. Since the level of synchrotron radiation is expected to be small even compared to a 1 μK CMBR anisotropy in the spectral region being considered, we can tolerate the substantial uncertainties inherent in this model in extrapolating from radio measurements down to sub-millimeter wavelengths. However, in the era of CMBR detections, all possible sources of contamination must be carefully excluded.

Foreground bremsstrahlung (free-free emission) poses a more worrisome contamination. Across the spectral region $\sim 5 - 10$ cm^{-1}, the spectrum of bremsstrahlung is nearly degenerate with the CMBR. Due to the flat spectrum and possibly strong self-absorption of bremsstrahlung at lower frequencies, radio surveys cannot provide a reliable means for subtracting this foreground. A reasonable estimate of the sky brightness from Galactic bremsstrahlung was obtained from the *COBE*/DMR maps by spectral decomposition after synchrotron and dust subtraction (Bennett *et al.* 1992). Based on this estimate, we again expect the amplitude of this foreground to be small, for our spectral bands, compared to the CMBR anisotropy. An important caveat here is that the DMR result smooths the sky with a 7° beam. If there is substantially more contrast at 0.5°, then bremsstrahlung can become more of a problem. We will address this question directly with the MSAM II experiment (Section 5.2).

3.1 TopHat Bands

To determine the CMBR anisotropy sensitivity for an experiment, we used a least-squares analysis for fitting the data to a series of alternative sky models. The channel placements shown in Figure 1 are optimized for these reasonable foreground models and use measured or projected instrument sensitivity.

The constraints on the band selections are: (a) no more than 5 spectral channels (due to geometry constraints within the dewar), (b) no band below ~ 5 cm^{-1} (to limit diffraction), and (c) no band at 18.5 or 25 cm^{-1} (to avoid strong H_2O atmospheric emission lines). Bandwidths of 1.5 cm^{-1} are chosen as a compromise between spectral resolution and signal sensitivity. Detector noise is modeled using a radiation-loaded detector model (Mather 1982, 1984a, 1984b) in addition to photon noise. Optical efficiency, conservatively estimated to be 15% at 5 cm^{-1}, tapering to 5% at 20 cm^{-1}, is based on our experience with the FIRS/MSAM I radiometer. Sky sensitivities assume 10^4 seconds of integration time per differential patch, allowing over 40 patches to be observed in 7 days with a 67% duty cycle. Sky models consist of the CMBR anisotropy spectrum, one parameter (amplitude only) or two parameter (amplitude and spectral index) dust emission, bremsstrahlung, and synchrotron radiation (Weiss 1980). Using contour plots of the CMBR sensitivity as a function of the band positions for each of the models considered, we found the configuration that is nearly optimal for all sky models and that avoids combinations with enhanced degeneracies (high correlations) between components. We place the TopHat radiometer bands at 5.5, 7, 8.5, 13, and 20 cm^{-1}. The analysis also reveals that it is possible to do almost as well with only four channels in roughly the same frequency range. However, the extra degree of freedom will increase confidence that the selected model is correct, as well as minimizing the risk that the actual sky spectrum corresponds to a high correlation case.

The CMBR sensitivities for this channel selection are presented for several sky models in Table 1 under the "TopHat" column. Note that the CMBR sensitivity becomes $\sim$ 5 times worse if bremsstrahlung or synchrotron emission are included in the decomposition. This is due to the near-degeneracy between bremsstrahlung and the CMBR near the region of the CMBR peak. While seeking a CMBR first detection, it may be reasonable to ignore these possibly small sources of foreground confusion. In the era of CMBR anisotropy detections, however, all sources of foreground must be considered, and we must pay special attention to estimating total measurement accuracy in the presence of these sources. Table 1 demonstrates a potential need for a low frequency companion to the TopHat experiment to remove the low frequency foregrounds in a manner similar to MSAM I/II combination.

Table 1. CMBR sensitivity (μK RMS) of the experiments for different spectral decompositions.

Model[d]	MSAM[a]			TopHat[b]	All-Sky Surveys[c]		
	I[e]	II	combined		FIRS[f]	COBE[g]	combined
CMBR	21.	3.2	3.2	0.68	100.	80.	62.
CMBR+Dust$_1$	22.	6.0	3.3	0.74	120.	82.	67.
CMBR+Dust$_2$	35.	44.[h]	5.8	1.2	210.	180.	82.
CMBR+Dust$_1$+Brem	240.	29.[h]	8.3	6.7	1500.	140.	95.
CMBR+Dust$_2$+Brem	240.[h]	—	15.	6.7	2600.[h]	360.	130.
CMBR+Dust$_1$+Brem+Sync	300.[h]	—	39.	7.5	4300.[h]	360.	170.

[a]Sensitivity per single difference, integrated for 10^3 seconds.
[b]Sensitivity per single difference, integrated for 10^4 seconds.
[c]Sensitivity per 2.5° pixel.
[d]Components are CMBR anisotropy, one- or two-parameter dust, bremsstrahlung, and synchrotron emission. See the text.
[e]Based on 20-minute integrated noise measured during June 1992 flight.
[f]Based on noise measured during October 1988 flight.
[g]Based on the DMR first-year map sensitivity. FIRAS used for dust removal.
[h]No extra spectral degrees of freedom—simple solution only.

3.2 Sensitivity of FIRS and MSAM

Table 1 also lists the CMBR sensitivity of the FIRS and MSAM experiments for the same foreground models. The columns for the FIRS/MSAM I radiometer and for DMR plus FIRAS, labeled *COBE*, are based on actual noise measurements. The MSAM II sensitivity is based on the estimates of Table 2 (see Section 5.2).

The *"combined"* columns give the result of a combined spectral decomposition of the same patch of sky using the instruments of the preceding two columns.

4 Far-Infrared Survey (FIRS)

The large-scale, Far-Infrared Survey (FIRS) experiment has been flown two times to produce a map of nearly one half of the sky with CMBR sensitivity similar to the *COBE*/DMR instrument but at frequencies near and above the CMBR peak at 6 cm^{-1}. The motivations for this experiment are summarized below.

1. The FIRS maps have an angular resolution of 3.8° so that a wide range of spatial scales outside the horizon (at the surface of last scattering) are observed. This optimizes the measurement of the initial, unevolved, CMBR fluctuation spectrum. The spectral index of these fluctuations can be constrained with the FIRS better than with DMR alone. In addition, the completion of the all-sky maps will permit us to make an independent measurement of the CMBR quadrupole.

2. The FIRS will greatly enhance the separation of CMBR fluctuations from foregrounds when combined with the DMR maps due to the enhanced spectral coverage. Two examples of source ambiguity in the DMR data alone are the possibility of Sunyaev-Zeldovich fluctuations (Hogan 1992) or a cold Galactic dust component, both of which can be detected far better with the combination of maps than with either map alone.

3. While both *COBE*/FIRAS and *COBE*/DIRBE cover this range with precise, absolute measurements, their sensitivity to fluctuations is not as high as the FIRS. As seen in Table 1, the CMBR sensitivity with foreground models which include more than a one parameter dust model or a single low frequency foreground are improved by a factor of two with the inclusion of the FIRS maps.

4. The completion of the FIRS will map the high Galactic latitude dust over the whole sky. A better removal of Galactic dust contamination (see Table 1) will be possible for all the DMR data.

5. The maps produced by the FIRS are completely independent of the DMR maps and have a different heritage. Unforeseen or undetected errors in either experiment could produce a systematic problem which could lead to false detections. The strong cross-correlation between the maps is the best evidence that CMBR anisotropy has indeed been detected (Ganga *et al.* 1993a)

Coverage of ∼ 90% of the entire sky will require three new flights of the instrument, one from the Northern hemisphere, and two from the Southern hemisphere. The first flight from the North 1) ensures that the instrument is functioning as it did in 1990, 2) covers the 20% of the Northern sky which remains unmapped, and 3) verifies that the overlapping regions produce maps that are consistent between flights.

4.1 Instrument

The FIRS bolometric radiometer has spectral bands centered at 5.7, 9.2, 16.5, and 23 cm^{-1} , each about 1 cm^{-1} wide, as shown in Figure 1 (Page *et al.* 1993). Measurements on the sky are referenced to an internal, temperature controlled load using a cryogenic chopper (Page *et al.* 1992) rather than the two beam DMR configuration. The measured, in-flight, sensitivity of this configuration is 600 μK$_{Planck}$ $\sqrt{sec}$ on the sky in each of the first two channels.

4.2 Results

The October 1989 flight has led to limits on the amplitude of CMBR fluctuations on 4° to 20° angular scales (Meyer *et al.* 1991, Ganga *et al.* 1993b). Detected structure in the sky rather than instrument noise led to this limit but the nature of the source could not be identified with the completed analysis of only one channel (5.7 cm^{-1}). This structure is now verified by the DMR to be CMBR anisotropy (Ganga *et al.* 1993a). We have, together with Richard Bond at CITA, reanalysed the map to constrain parameters of structure evolution models (Bond *et al.* 1993).

Data from the May 1990 flight are currently being added (Ganga *et al.* 1993c). The next analysis step is to combine these maps with those from the higher frequency channels to provide strong spectral constraints on the thermal nature of the detected signal.

5 Medium-Scale Anisotropy Measurement (MSAM)

The goals of the Medium-Scale Anisotropy Measurement (MSAM) project are three-fold; 1) to complete a set of measurements of CMBR anisotropy at the $\delta T/T \sim 10^{-5}$ level, 2) to verify that any detections are not a result of systematic errors or foreground emission sources, and 3) to develop and test TopHat instrumentation. We will accomplish these goals by reflying the current MSAM I configuration, and subsequently flying the long-wavelength radiometer (MSAM II), appropriately upgrading the gondola with prototype TopHat hardware.

5.1 Repeat Flight of Existing System (MSAM I)

The existing system consists of the radiometer used in the FIRS coupled with a 1.5 meter, pointing controlled, balloon-borne telescope built at GSFC. The instrument has a 28′ beam and a three position secondary chopper, which throws the beam 1.7° peak-to-peak. This configuration allows for sampling two distinct scales of spatial separation (see the next section for a more complete discussion).

The first MSAM flight, in June 1992, produced 4.7 hours of CMBR anisotropy data in a circumpolar region. Preliminary results (Cheng *et al.* 1993a) from this flight have provided a detection of $\delta T/T \sim 2 \times 10^{-5}$. To verify this detection, we will refly the instrument to observe the same region of the sky.

Chopping Strategy For a sky difference measurement, the magnitude of atmospheric contamination decreases as the chopper throw decreases. Unfortunately, on ~ 1° angular scales, the predicted sensitivity to the CMBR fluctuation spectrum also decreases with decreasing chopper throw (Gorski 1993, Bond and Efstathiou 1987). Thus, there is a tradeoff which can only be optimized during a flight because the detailed sky conditions are not known beforehand (and may in fact be variable). The MSAM/TopHat chopper is a driven, three-position, mirror mechanism which cycles the radiometer input to a left (L), center (C) and right (R) beam on the sky. This information is sampled and returned to the ground so that two *independent* signals can be simultaneously extracted. The double difference ($C - L + C - R$) has a characteristic separation of half the throw distance, while the single difference ($L - R$), has twice this characteristic separation. The sky noise difference in these two signals (after accounting for the integration times) is a measure of the potential atmospheric contamination. The chopper allows for a reasonable range of adjustment in the throw (0° to 1.7° for MSAM, 0° to 3° for TopHat), which can be commanded to the optimum value. Thus, while observing at a particular throw angle, we can sample the sky at two angular scales, both for characterizing the CMBR and for monitoring any time-variable atmospheric noise. The throw angle can be set to the maximum value which is not contaminated by atmospheric noise. The existing MSAM secondary chopper has performed flawlessly both in ground tests and in flight.

5.2 Two Flights of the Long-Wavelength ADR Radiometer (MSAM II)

We will make two additional flights with the MSAM gondola, using a new radiometer with spectral channels designed to complement the MSAM I instrument. This lower-frequency system is more sensitive (see Table 2), and uses single-mode optics to increase sidelobe rejection. The combination of low and high frequency measurements on the same region of sky allows us to establish whether bremsstrahlung-like radiation or extragalactic sources are significant for CMBR measurements on 0.5° scales.

Table 2. MSAM II Radiometer Sensitivities

Frequency Range	60-80	80-110	130-170	GHz
Optical Efficiency	0.3	0.3	0.3	
Background Noise (BLIP)	0.7	1.1	1.2	10^{-17} W/$\sqrt{\text{Hz}}$
Detector Sensitivity[a]	0.5	0.5	0.5	10^{-17} W/$\sqrt{\text{Hz}}$
Receiver Sensitivity[b]	170	170	190	μK $\sqrt{\text{Sec}}$

[a]With radiative loading from sky, atmosphere, and warm optics.
[b]Sensitivity to a 2.7K Planck anisotropy spectrum, single difference.

The new instrument observes in three bands, centered at 70 GHz, 100 GHz, and 150 GHz, using monolithic bolometers cooled to 0.1 K by an ADR. This instrument is under development at Brown University and will be integrated with the existing MSAM I telescope. We are investigating the possibility of splitting the highest frequency band in order to improve spectral discrimination.

Optics The new radiometer will sample the sky with the same beam size used on all previous MSAM flights. At these longer wavelengths, the telescope is fed by single-mode optics ($A\Omega = \lambda^2$) to minimize diffraction effects. This approach has the additional advantage that the optical system can be designed accurately using Gaussian optics, and that well-developed microwave techniques can be used for dichroics, filtering, feedhorns, and coupling to the bolometers.

Since we want to observe the same spot on the sky simultaneously in all three wavelength bands, the existing chopping secondary mirror will be illuminated by three single-mode corrugated feedhorns through two dichroic plates. The feeds will have identical aperture diameters so that the primary will be illuminated with spot sizes that are inversely proportional to their wavelengths. The resulting diffraction-limited beam on the sky is a Gaussian with identical spatial width in each band.

Detectors The detectors are monolithic silicon bolometers coupled directly into a waveguide. At long wavelengths, this arrangement has two important advantages over free-space optics: high coupling efficiency and small size (smaller than a wavelength) so the detectors have intrinsically short time constants and a reduced cross-section to cosmic ray hits.

These detectors are fabricated at GSFC with a silicon micromachining process which defines the thermally isolating legs (Moseley *et al.* 1992). The thermistors are ion-implanted in silicon and are read out through ion-implanted electrical leads on the surface of the legs. Existing devices have been developed for a variety of missions from airborne (Kuiper) and satellite (AXAF) platforms. Electrical NEP sensitivities of 5×10^{-18} W/$\sqrt{\text{Hz}}$ at 70mK are routinely achieved. Optical sensitivities with radiative loading are expected to be similar.

We are also considering a hot electron bolometer (Nahum and Martinis 1993), which has demonstrated an electrical NEP of 3×10^{-18} W/$\sqrt{\text{Hz}}$ at 100 mK. These devices are ideally suited for the single-mode coupling scheme described here and, if further investigations show them to be advantageous, we will use them instead of the monolithic silicon bolometers.

Cryogenics The detectors and filters will be cooled to 100mK by an ADR optimized for ballooning (Timbie 1990). This refrigerator is a refinement of the design which was prototyped for *SIRTF*. Design and construction is complete and it is currently being tested at Brown University. This system will remain cold for 24 hours, making it unnecessary to recycle during a flight.

6 TopHat

Current intermediate angular scale ($\sim 1°$) CMBR anisotropy measurements are aimed at achieving sensitivities on the order of $\delta T/T \sim 10^{-5}$, a level which is tantalizingly close to many predictions based on the *COBE* detection of large-scale anisotropy. Indeed, there have been several recent reports of detections at these levels (Cheng *et al.* 1993a, Gaier *et al.* 1992, Fischer *et al.* 1992), but the source of the signals has yet to be unambiguously determined. Reliable measurements at these angular scales hold the unique promise of constraining initial conditions for theories of structure formation, but depend on some key experimental advances.

1. Control of systematic errors by proper design of the experimental configuration and observing strategy.
2. Demonstration of freedom from residual systematic effects by in-flight measurements.
3. Improved sensitivity to CMBR anisotropy signals in the presence of potential astrophysical foregrounds.

At the $\delta T/T \sim 10^{-6}$ level, balloon measurements, including the MSAM, suffer from severe geometric constraints. With the Earth below and the balloon above, only a limited angle is available where instrument sidelobes are not a serious concern. Observations are almost always made through about twice as much residual atmosphere as is otherwise necessary to ensure that the balloon and support structures, which are rotating above the gondola, do not affect the signals.

The advent of polar long duration ballooning, with flights lasting two weeks or more, has created the opportunity to make a relatively low-cost instrument that can be sufficiently tested under observing conditions to justify a factor of ten increase in instrument sensitivity.

We are constructing a one meter telescope and radiometer which can be placed on top of a balloon and flown from Antarctica for 2 weeks. Nearly half of the sky would be clear of Earth, Sun, Moon, balloon, and atmosphere for the entire two week flight. The instrument, protected behind a simple Sun-screen, is in a thermally stable environment. We can observe at elevation angles of 70° to reduce the effects of residual atmosphere. The ground shields needed for reflecting the telescope sidelobes to the cold sky are small because all of the significant emission sources are at large angles ($> 70°$) from the instrument beam. The radiometer has an intrinsic sensitivity of 3 times that of our existing instrument, so we need not extrapolate our current understanding of the atmosphere and systematic effects beyond what is reasonable. With one polar long duration flight of the TopHat instrument, we can observe 40 sky positions with a 1 μK RMS sensitivity for each point, including the uncertainty due to dust emission.

Approximately 1/2 of the observing time would be devoted to a study of stability and known systematic effects. It is worthwhile to emphasize that the time available for performing tests of systematic contamination is a critical factor favoring a long duration flight. Indeed, an important lesson from the *COBE* experience is that a longer observing period not only increases sensitivity, but allows for many of these tests to be run (Kogut *et al.* 1992). Without this capability, confidence in the final result will be jeopardized.

6.1 Payload Overview

The radiometer, telescope, radiation shield, Sun shield, telescope mount, and a portion of the electronics will be mounted on the balloon top plate (Figure 2). Because of the weight constraint, this is an unusual configuration for normal scientific payloads, but not for balloon engineering experiments. The batteries, solar panels, communications equipment, and data storage equipment will fly in the conventional location suspended below the balloon. Electrical cables, embedded in the balloon during its construction, will provide the power, control, and data links. The total top payload will weigh less than 140 pounds, the limit of what has been previously launched in this configuration. The bottom payload, including the communications and control electronics provided by the National Scientific Balloon Facility (NSBF), will weigh approximately 1000 pounds.

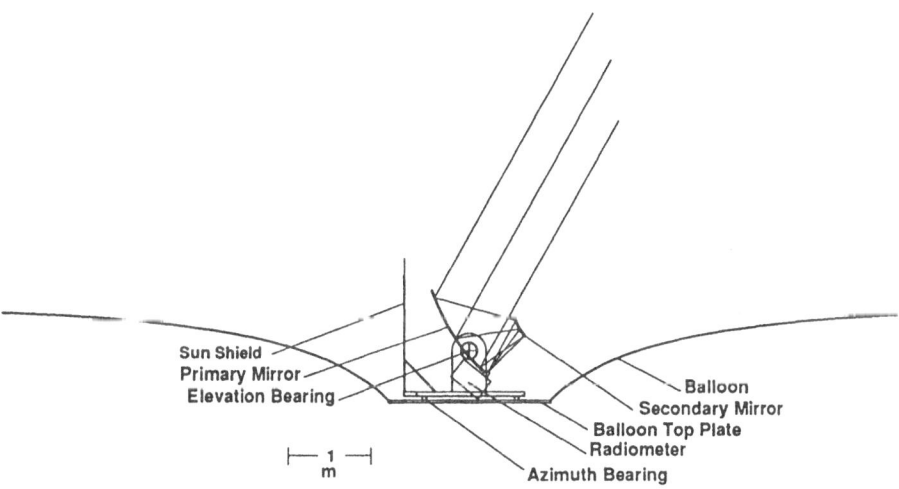

Fig. 2. Schematic of the TopHat Platform *in situ*. The ground screens are omitted for clarity.

The top structure will be made of carbon-fiber composite, yielding a projected top payload weight of approximately 100 pounds. The telescope will be in an altitude-azimuth mount, rotating relative to the balloon top plate on a servoed azimuth bearing. The pointing is controlled with a combination of rate gyro, Sun sensor, magnetometer, star camera, and local vertical reference. The magnetometer is included to permit test flights at mid-latitudes. This configuration is a natural adaptation of the existing MSAM gondola control system, with the inclusion of the local vertical and Sun sensors.

Conditions on the top of the balloon are not as fully understood as those for a suspended payload. There are modes of oscillation of the balloon in addition to the swinging that all payloads must suffer. On the other hand, the torsional coupling to the balloon is much tighter than in the hanging configuration, eliminating the need for a momentum wheel and a two-stage azimuth servo. We plan to measure the detailed motion of the top of the balloon with a special test flight of a top-mounted engineering package to verify the results of our modeling.

6.2 Telescope Optics

The telescope has a 30° off-axis, one-meter diameter primary mirror. The nutating secondary mirror is 170 mm and also 30° off-axis. The telescope is designed to observe at an elevation angle of approximately 70° to keep both the Earth and the Sun far from the main beam, and to minimize atmospheric noise. Only the central 50% of the primary is illuminated to minimize sidelobe response. The geometry of the optics is a simple scaling of the MSAM I configuration, enabling reliable predictions of far-sidelobe response based on existing measurements.

The design of the ground screen in this off-axis geometry is governed by the usual constraints. It is highly desirable to have all the mirror surfaces and edges view only the reflective ground screen or the sky. It is also important to avoid forming cavities, as their relatively high emissivity can be modulated by the secondary mirror to create systematic offsets. With the Sun, Moon, and planets all below 30° elevation, a very large solid angle is clear of sources for the entire duration of the measurement. A shallow open bucket with a single reflecting shield between the package and the Sun can provide complete protection from terrestrial and solar radiation while simultaneously avoiding the cavity problem.

6.3 Long Hold-Time Dewar

A key requirement of TopHat is a cryostat that remains cold for several weeks with no servicing. Placing the telescope on top of the balloon constrains the dewar weight. The total weight of the telescope is a strong function of the dewar weight because mechanical structures, bearings, and motors are sized to support the components on the elevation mount. We are in the process of constructing a ^{3}He evaporation cryostat which will weigh 25 pounds when full of cryogen and hold for 25 days at float altitude. The lowest resonance for this cryostat is designed to be above 80 Hz to minimize coupling with the servo system and the detectors. Design and construction are almost complete and the dewar will be operational by the end of 1993.

This dewar concept is derived from the ^{3}He refrigerator built for the FIRS and MSAM I measurements (Section 4.1, Cheng *et al.* 1993b). The FIRS/MSAM I dewar already holds at 0.24 K for 8 days so the performance of the new dewar is a reasonable extrapolation of existing technology.

7 Conclusions

Measurements of the CMBR anisotropy will be facing many new experimental challenges as we attempt to push sensitivity levels by yet another factor of ten. However, with the clear detection of large-scale anisotropy by *COBE* and FIRS, we have finally advanced from the discovery phase to one of measurement and characterization. That is, we now have at least a crude idea of the nature of the target. A deeper understanding of the nature of the CMBR anisotropy at these levels is a crucial step to the ultimate goal of mapping the CMBR at higher resolution and over large portions of the sky.

References

Bennett, C. *et al.* 1992, *ApJ*, **396**, L7.

Bond, J.R., Efstathiou,G. 1987, MNRAS, **226**, 655.

Bond, D., Cheng, E.S., Cottingham, D.A., Meyer, S.S., Page, L.A. 1993, *ApJ*, (in press).

Brandt, W.N., Lawrence, C.R., Readhead, A.C.S., Pakianathan, J.N., Fiola, T.M. 1993, (preprint).

Cheng, E. S., *et al.* 1993, *ApJ*, (submitted 15 May 1993).

Cheng, E.S., Meyer, S.S., Page, L.A. 1993, R.S.I. (in preparation).

Efstathiou, G., Bond, J.R., and White, S.D.M., 1992, MNRAS, **258**, 1.

Fischer M. *et al.* 1992, *ApJ*, **388**, 242.

Ganga, K., Cheng, E.S. , Meyer, S.S., and Page, L.A. 1993, *ApJ*, **410**, L57.

Ganga, K., Bond, R., Cheng, E.S., Meyer, S.S., and Page, L.A. 1993, APS Meeting, Washington, D.C.

Ganga, K., Cheng, E.S., Cottingham, D.A., Meyer, S.S., and Page, L.A. 1993, *ApJ*, (in press).

Gaier, T., Schuster, J., Gundersen, J., Koch, T., Seiffert, M., Meinhold, P., Lubin, P. 1992, *ApJ*, **398**, L1.

Gorski K. 1993, *ApJ*, (in press).

Gundersen, J.O. *et al.* 1993, Center for Particle Astrophysics Preprint, Berkeley, CA.

Hogan, C.J. 1992, (preprint).

Kashlinsky, A. 1992, *ApJ*, **387**, L1.

Kogut, A. *et al.* 1992, *ApJ*, **401**,1.

Mather, J.C. 1982, *Appl. Optics*, **21**, 1125, 1984, *Appl. Optics*, **23**, 584, and 1984, *Appl. Optics*, **23**, 3181.

Meinhold P., and Lubin, P. 1991, *ApJ*, **370**, L11.

Meinhold P., Devlin, M., Fischer, M., Gundersen, J., Holmes, W., Lange, A., Lubin, P., Richards, P., Smoot, G. 1993, *ApJ*, (accepted).

Meyer, S.S., Cheng, E.S., and Page, L.A. 1991, *ApJ*, **371**, L7.

Moseley, S. H., *et al.* 1992, Proc. ESA Symp on Photon Detectors for Space Instrumentation, ESA-SP-356, p 13 1992

Myers, S.T., Readhead, A.C.S., and Lawrence, C.R. 1992, *ApJ*, (In Press).

Nahum, M., and Martinis, J. M., 1993, "Novel Hot-Electron Bolometer", to appear in proceedings 20th conference on Low-Temperature Physics, Eugene, OR.

Page, L.A., Cheng, E.S., Meyer, S.S. 1990, *ApJ*, **355**, L1.

Page, L.A., Cheng, E.S.,and Meyer 1992, *Appl. Optics*, **31**, 95.

Page, L., Cheng, E., Golobovic, B., Gundersen, J., and Meyer, S. 1993, *Appl. Optics*, (accepted 20 April).

Readhead, A.C.S., Lawrence, C.R., Myers, S.T., Sargent, W.L.W., Hardebeck, H.E. 1989 , *ApJ*, **346**, 566.

Smoot *et al.* 1992, *ApJ*, **396**, L1.

Timbie, P.T., Bernstein, G.M., and Richards, P.L. 1990, *Cryogenics*, **30**, 271.

Watson, R.A., Gutierrez de la Cruz, C.M., Davies, R.D., Lassenby, A.N., Rebolo, R., Beckman, J.E., and Hancock, S., 1992, *Nature*, **53**, 660.

Weiss, R. 1980, *ARA&A*, **18**, 489.

Wright *et al.* 1991, *ApJ*, **381**, 200.

The Current Status of the Tenerife Experiments and Prospects for the Future.

A.N. Lasenby[1], R.D. Davies[2], S. Hancock[1], C.M. Gutierrez de la Cruz[3],
R. Rebolo[3] and R.A. Watson[2]

[1] Mullard Radio Astronomy Observatory, Cavendish Laboratory, Madingley Road, Cambridge CB3 OHE, U.K.
[2] University of Manchester, Nuffield Radio Astronomy Laboratories, Jodrell Bank, Macclesfield SK11 9DL, U.K.
[3] Instituto de Astrofisica de Canarias, 38071 La Laguna, Tenerife, Spain.

1 Introduction.

The Tenerife experiments are directed towards the detection of cosmic microwave background (CMB) fluctuations on scales of a few degrees. Such fluctuations at rms levels between $\Delta T/T = 3 - 30 \times 10^{-6}$ are predicted by most scenarios of galaxy formation; they are impressed by the Sachs-Wolfe effect on the 2.7 K flux emerging from the recombination process at z=1000. Degree scales are unique in providing direct information about the intrinsic CMB era, in contrast to arcminute scales which are potentially referring to recombination or the Sunyaev-Zeldovich effect in the foreground at lower redshifts. In addition, observations on angles greater than the horizon scale ($\sim 2°$) are causally unconnected on standard hot Big Bang models and allow tests of inflationary scenarios for explaining structure in the early Universe. It is interesting to note that as surveys for structure in the early Universe reach greater distances, networks and associated voids on 100 Mpc scales (corresponding to 1° at z=1000) have been identified.

All searches for CMB structure must take serious account of foreground radiation, particularly that from the Galaxy. We have chosen to make our searches at the lower frequencies where synchrotron radiation is likely to be the contaminant rather than higher frequencies where thermal emission from interstellar dust will dominate the foreground; it is our belief that synchrotron (and free-free) emission are more thoroughly understood and have less free parameters. Accordingly our radiometers have been operated at 10.45, 14.9 and 33 GHz giving CMB structure information covering the angular range $5° - 15°$. We will show in this article that at 33 GHz the Galactic contribution to the observed signals is negligible. The Teide Observatory sited on Tenerife is found to be suitable for efficient CMB observations at frequencies up to 33 GHz.

We will describe the extension of the present programme, which has successfully detected structure on a coherence scale of 4°, to scales in the range 10 arcmin to 2°, where many scenarios of star formation predict a maximum in the fluctuation amplitude. This includes a two-element interferometer at 33GHz, a three-element interferometer sited at Cambridge, called CAT, and the proposal for the Very Small Array (VSA).

2 Realisation of the Beam-Switching Radiometers.

At radio frequencies the weak signals expected from the CMB structure require long integration times with sensitive receivers. The beam-switching technique reduces the effects of long-term changes in the atmosphere and instrumental drifts, thus enabling long integrations to be made. The penalty paid for switching is an increase in effective rms receiver noise of a factor of $\sqrt{2}$ for single switching and of 2 for double switching as employed in our experiments.

The experimental arrangement is shown in Fig. 1. Fast switching (32 Hz) between two corrugated horns is combined with a 7 sec wagging motion of the plane

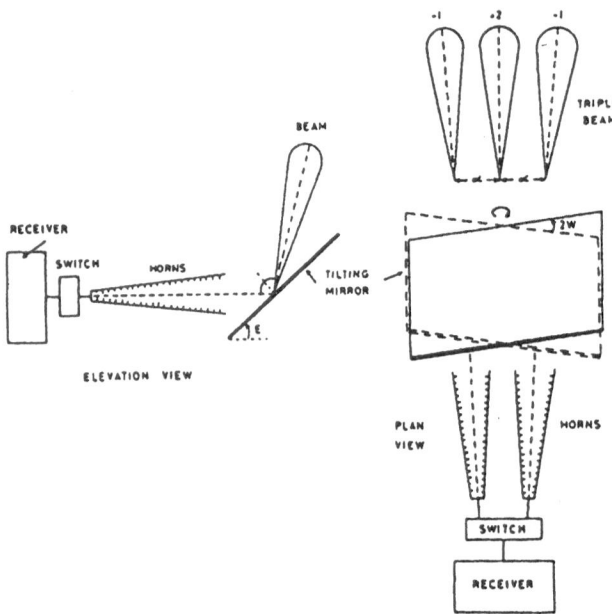

Fig. 1. The operation of the beam-switching radiometer to provide a triple beam response through a combination of fast switching between two horns and a wagging mirror.

mirror. The declination of observation is determined by adjustment of the elevation angle of the wagging mirror. The experimental details are given in Davies *et al.* [1]. A feature of our system is the use of a 4-port ferrite circulator switch to allow two independent output channels. The resulting Gaussian beamwidth is 5.°5 FWHP with reference beams of the same width displaced ±8.°2 in azimuth (right ascension). The receiving system is fixed on the ground and scans a fixed declination by Earth rotation.

Three experiments are operated at Teide Observatory, all with the same beamwidth so that direct comparisons of spectral structure can be made. The frequencies chosen are 10.45, 14.9 and 33 GHz for which the theoretical sensitivities are 5.5, 3.5 and 2.1 mK IIz$^{-\frac{1}{2}}$, based on $T_{sys}/\sqrt{\Delta\nu \cdot \tau}$. Data at these frequencies have been taken over a number of years. The data at a particular frequency and declination for each day of observation are edited and stacked to give the final high sensitivity scan. Editing involves the elimination of data taken within 50° and 30° of the Sun and Moon respectively and of data where atmospheric effects increased the rms noise in the 80 s integration cycle by more than a factor 2.

3 Data Processing.

The data reported here are for observations of a strip of the sky covering 100° in right ascension, at a declination of +40°. This area was selected after consideration of foreground emission from the Galaxy and discrete radio sources and defines a window over RA 161° − 260° through which effective CMB observations are possible. By means of repeated drift scan observations over recent years, useful integration times have been built up for this region at 10, 15 and 33 GIIz thereby providing high sensitivity to CMB features (see contribution of Watson et al. , this volume).

For each scan the data stream, consisting of successive 80s integrations is binned into 4 minute cells, before removal of non-astronomical baselevel drifts, as derived from an MEM sky brightness solution (Hancock et al. in preparation). Subsequent weighted combination of the daily drift scans provides a final stacked data scan with a consequent improvement in the signal-to-noise ratio. The 1° binned data are oversampled and further improvement in the signal to noise to beam sized features (for which the experiments have peak sensitivity) is facilitated by binning at the Nyquist rate for structures of a beam width, i.e. at 4° intervals. The 4° binned data points at 10, 15 and 33 GHz are depicted in figure 2 together with confidence limits at the 68 % level, as derived from the scatter over the individual scans. Structure is visible in each data set and the reality of this is supported [3] by both a χ^2 analysis and the analysis of independent subsets of the data.

Despite the apriori selection of the optimum sky area for observation, it is expected that discrete radio sources and Galactic emission may be a contributor to the signals seen in the data. Indeed from estimates of the expected point source component at each frequency, as constructed from the Kuhr catalogue [2], it is clear that with the existing scan sensitivity the source 3C345 should be clearly visible in the 10 and 15 GIIz data, centred on RA 250°. This is seen to be the case, as is evidenced by the dotted lines in figure 2, which represent the discrete source contribution after convolution in the instrument beams. That the structures seen at RA 250° in the independent 10 and 15 GIIz experiments agree so well with the estimates implies that:

- i) the experiments consistently detect known structures
- ii) the structure present over RA 161° − 230° is probably real and cannot be principally due to discrete source emission

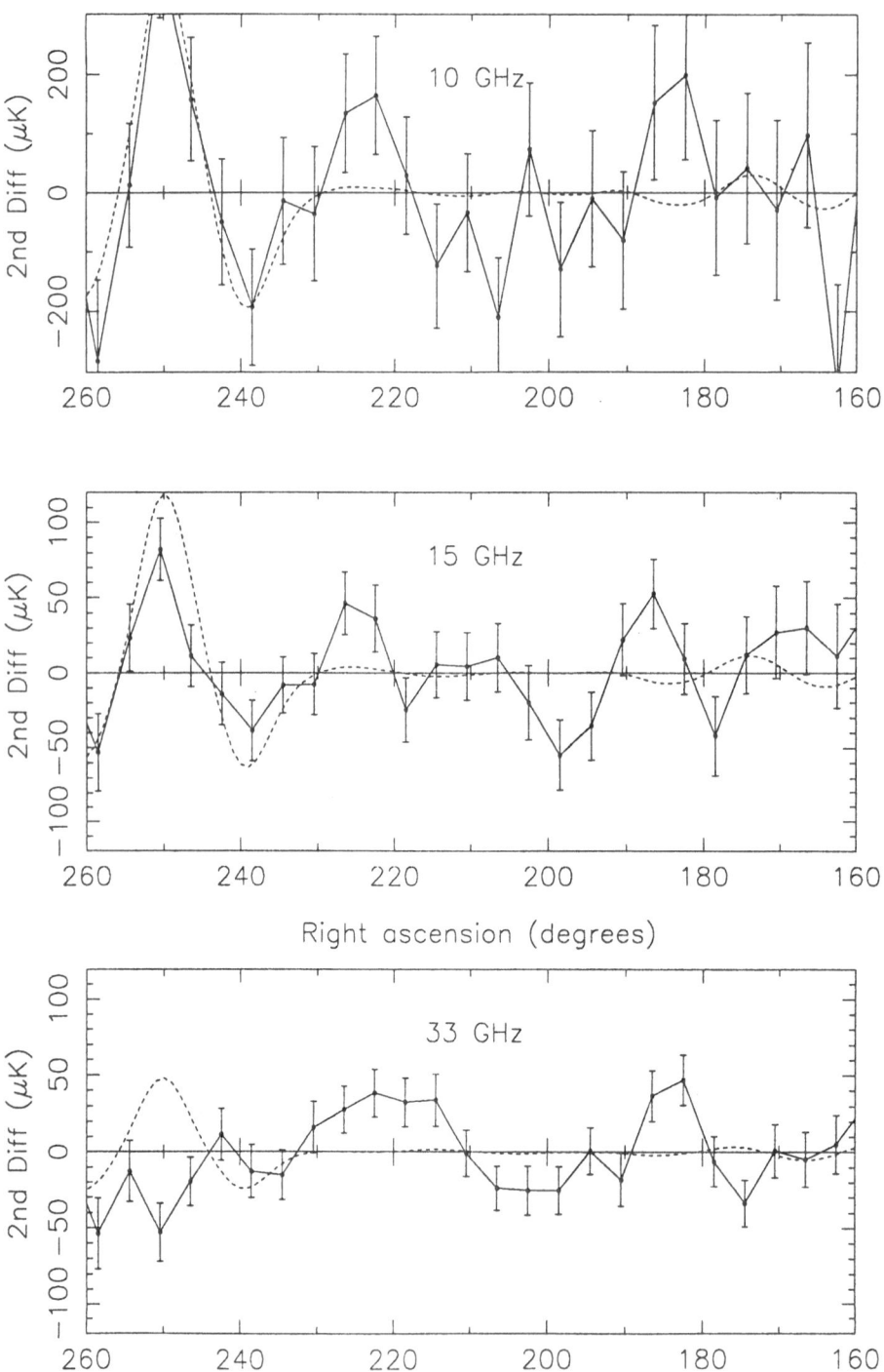

Fig. 2. Final stacked data scans at 10, 15 and 33 GHz. The dotted lines represent the estimated contribution from discrete radio sources.

– iii) the absence of the radio source in the 33 GHz data is not unexpected if its spectrum steepens at higher frequencies, as is likely.

We proceed to concentrate on the region RA 161° − 230°, making the minor correction for the estimated discrete source emission over this restricted range.

4 Statistical Analysis of Data.

Well-defined features are apparent in the data scans and various statistical tests have been applied to the data [3] to determine the significance and amplitude of such structures. The benchmark for this type of analysis is the likelihood analysis assuming a Gaussian model for the intrinsic sky auto-correlation function:

$$C_{intr}(\theta) = <\Delta T(\mathbf{n})\Delta T(\mathbf{n}') > = C_o \exp\left(\frac{-\theta^2}{2\theta_c^2}\right) \tag{1}$$

for angles θ subtending $\mathbf{n}$ and $\mathbf{n}'$ on the sky. Whilst being entirely non-physical, this sky model is simply described by two parameters $\sqrt{C_o}$, the rms fluctuation amplitude and θ_c, the coherence scale of the fluctuations. It is also the form conventionally quoted by observers and is useful to compare with the results from other experiments.

One can form the likelihood function [2] for a range of $\sqrt{C_o}$ and θ_c. The Tenerife configuration attains peak sensitivity to intrinsic sky structures with $\theta_c \sim 4°$ and thus it is necessary only to vary $\sqrt{C_o}$. The favoured model value is defined by the peak in likelihood and a significant peak for non-zero $\sqrt{C_o}$ constitutes a detection of structure at that amplitude. Figure 3 shows the likelihood function for the 33 GHz data and is characteristic of a detection of structure at $60 \, ^{+16}_{-16} \, \mu K$. Equivalent plots at 15 and 10GHz give detections at $\sqrt{C_o} = 48 \, ^{+16}_{-16} \, \mu K$ and $\sqrt{C_o} = 29 \, ^{+20}_{-30} \, \mu K$ respectively [3]. A detailed analysis using a more physical power law model for the power spectrum of the fluctuations supports the multi-frequency detection of structure.

The component of the structure responsible for the detections is seen to correlate between the independent scans [3] and from this one can infer that the structure has a common origin, this being either Galactic synchrotron/free-free emission or CMB. Discrimination between these candidates is based on their spectra, with synchrotron and free-free obeying $T_A \propto \nu^{-\beta}$, with β typically between 2.0 and 3.0, and CMB having a Planckian spectrum, that is $T_A \propto \nu^0$ in this spectral region. The fall in amplitude by a factor 30 for synchrotron and 10 for free-free between 10.4 and 33 GHz, is not observed, whereas the data are in agreement with a blackbody emission spectrum.

The conclusions to be drawn from the above are thus that the Tenerife data exhibit statistically significant detections of CMB structure, which for a Gaussian model auto-correlation function have an rms amplitude of $54 \, ^{+14}_{-10} \, \mu K$ (for a weighted sum of 15 and 33 GHz). That these detections can now be associated with clearly defined CMB structures offers a useful advance over the published

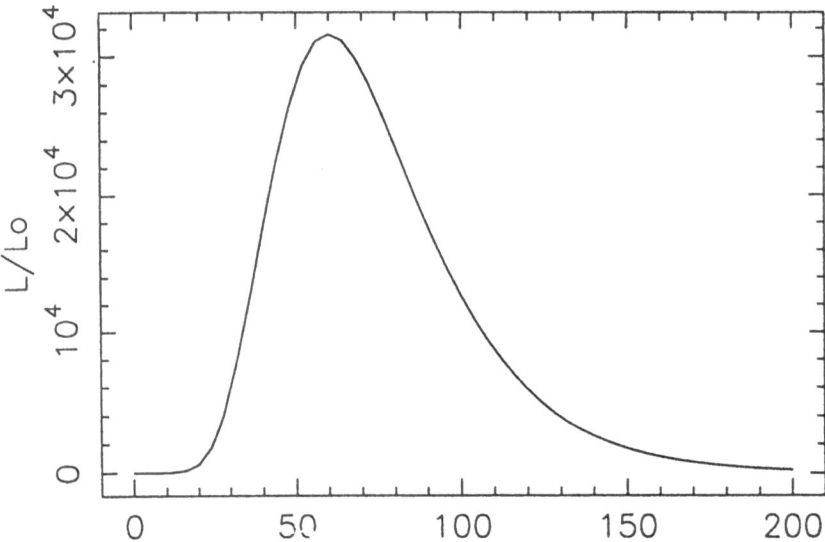

Fig. 3. The likelihood function for the 33 GHz data assuming a coherence scale of 4°.

COBE results [4]. The statistical amplitude of the Tenerife structure is entirely consistent with the COBE result [3] and can be used both to extend the information on the fluctuation power spectrum from $\theta \gtrsim 10°$ down to $\theta \gtrsim 4°$ and to place additional constraints on the spectral index n giving $n > 0.9$ as a lower bound at 68 % confidence.

5 Future Projects.

Improvements in the large scale ($\gtrsim 2°$) CMB anisotropy results are expected in the coming year, with the extension of the Tenerife sky coverage by a factor of 5 and also the forthcoming results from the COBE DMR second year of observing. With the improvements in the DMR sensitivity to features it should be possible to compare maps between Tenerife and DMR and as a consequence to formulate a powerful test of inflationary models. Whilst the importance of such large-scale observations needs no explanation, the case for observations on smaller scales is also clear, since these fluctuations correspond to mass scales visible in the present day Universe. As yet, the situation regarding the detection of CMB anisotropy on scales from 1' to 2° remains uncertain. It is hoped that the initiation of three new ground based interferometers will help to resolve the issue and allow further mapping of the fluctuations.

CAT.

The Cosmic Anisotropy Telescope [5] utilises a unique three element interferometer arrangement to observe the CMB on scales $\sim 0.5°$ where galaxy formation scenarios predict peak power. The instrument has recently begun observations at 13–17 GHz from a sea level site at Cambridge, U.K. Despite the relatively poor atmospheric conditions at Cambridge, the interferometric rejection of the atmosphere allows a projected sensitivity of $\lesssim 40$ μK per week. The instrument should thus have the capacity to unambiguously probe CDM galaxy formation scenarios.

33 GHz Interferometer.

A short-baseline interferometer operating at 33 GHz is being constructed at Jodrell Bank for observations at Teide Observatory in 1994. The main-lobe response at half-power is $2^\circ.5$ in declination and $2^\circ.0$ in right ascension. With a 10 percent bandwidth, the expected sensitivity is 0.6 mK Hz$^{-\frac{1}{2}}$. Experience with the 33 GHz beam-switching radiometer at Teide Observatory indicates that a large fraction of the time will be suitable for interferometry. Furthermore, extensive observations with a 5 GHz short-baseline interferometer at Jodrell Bank have measured galactic and extragalactic confusion levels; when extrapolated to 33 GHz these lead to less than 5 μK of background confusion.

Very Small Array.

Both CAT and the 33 GHz interferometer are restricted to scales $\sim 0.5^\circ$ and due to the limited number of Fourier components sampled are not capable of detailed imaging. The role of the VSA is to provide just such imaging over the angular range $10'$ to 2°. The proposed instrument design incorporates 10–15 horn elements observing over a 10 GHz bandwidth, in 4 frequency channels centred on 33 GHz. The variety of baselines will allow detailed mapping and the range of observing frequencies should provide the necessary spectral information. It is proposed to locate the instrument at Teide Observatory in Tenerife, where the combination of the interferometric technique and the good observing site should allow the target sensitivity of $\sim 6\mu$K to be reached in a few months.

References

1. Davies, R.D., Watson, R.A., Daintree, E.J., Hopkins, J., Lasenby, A. N., Sanchez-Almeida, J., Beckman, J.E. & Rebolo, R. Mon. Not. R. astr. Soc., 258, 605 (1992).
2. Watson, R.A., Gutierrez de la Cruz, C.M., Davies, R.D., Lasenby, A.N., Rebolo, R., Beckman, J.E., & Hancock, S. Nature, 357, 660 (1992).
3. Hancock S., Davies, R.D., Lasenby, A.N., Gutierrez de la Cruz, C.M., Watson, R.A., Rebolo, R. & Beckman, J.E. submitted to Nature.
4. Smoot, G.F. et al. Ap. J., 396, L1 (1992).
5. Robson M., Yassin, G., Woan, G., Wilson, D.M.A., Scott, P.F., Lasenby, A.N., Kenderdine, S. & Duffett-Smith, P.J. Astron. Astrophys., 277, 314 (1993).

Making Maps with the Tenerife Data

Robert Watson[1][2], *Rafael Rebolo*[1], *Carlos Gutierrez de la Cruz*[2], *Stephen Hancock*[3], *Anthony Lasenby*[3] *and Rod Davies*[2]

[1] Instituto de Astrofisica de Canarias, C/ Vía Láctea s/n, 38200 La Laguna, Tenerife, SPAIN.
[2] Nuffield Radio Astronomy Laboratories, Jodrell Bank, Macclesfield, U.K.
[3] Mullard Radio Astronomy Observatory, Cavendish Laboratories, Cambridge, U.K.

1 Introduction

Our set of switched-beam radiometers, known collectively as the Tenerife experiment, have an observing scheme consisting of a set five declination scans at $\delta = +35°, +37.5°, +40°, +42.5°$ and $+45°$, in which the highest priority was given to the centre declination. To date all our reported results have been made using this deeper surveyed area as the other scans were not capable of supplying any significant extra information. Now, with strong indications of features appearing in the latest processed scans (see the contribution of Lasenby et al. in these proceedings), the best strategy would appear to be to consolidate observations on the less sensitive adjacent scans. The outcome of this will be the production of a map, albeit an elongated one. Here we present the first limited maps at 10 and 15GHz and make tentative identifications. Also with recent improvements in sensitivity we review the possiblities for a finalized map at all three frequencies, which would provide a valuable extension to the COBE maps over our observed area at higher resolution and sensitivity.

The success of the COBE maps (Smoot et al 1992) has amply demonstrated the usefulness of a direct graphical representation of the "last scattering surface". This neatly bypasses the problems of choosing appropriate statistical models (Watson and Gutierrez de la Cruz 1993) as the true one is contained in the map. The resulting analysis of maps is very open-ended and yields to numerous approaches and has even created new methods (eg PIP analysis) and brought in techniques from other fields (eg pattern recognition), some of which are reported in these proceedings.

To make a meaningful map, most of its pixels should contain some tangible signal. For the weak signals the perturbations offer, this requires extensive integration times. Present radiometer systems require days to reach the required level, so an entire map consisting of several hundred such fields would take years. This is why experiments of the past have only consisted of a dozen points or so, because of the time limitations imposed by balloon flights or the weather on ground-based experiments.

2 Comparing Tenerife and COBE Sensitivities

discuss what information can be expected from a ground-based experiment such as Tenerife as compared to that already found by the COBE satellite. Table 1 shows the observing sensitivities of the Tenerife and COBE radiometers. The Tenerife receivers reflect the sensitivity of systems now available as opposed to the older technology "frozen in" at the design stage of the satellite. Weather and limited elevation range restrict both the available observing time and the sky coverage and therefore observations have to be made over a confined region. Curiously, the best sensitivities per beam area of the two experiments are similar as also in Table 1. The total area covered by the five declination bands is about 10% of the whole sky made up of approximately 200 independent beams of 5° FWHM. For comparision the COBE maps consist of 1000 or so fields of 7° FWHM covering the whole sky. Assuming the usable data from the 10, 15 and 33GHz radiometers are 50%, 25% and 15% respectively of the total time available and that each declination band is sampled uniformly, we can estimate the noise level per beam we should achieve in one year. These values are shown in the bottom row of Table 1 together with the same estimates for COBE, showing that good sensitivity should be possible if both the weather and instruments behave as assumed. In conclusion the Tenerife experiment can certainly compliment the information obtained from the COBE maps through higher resolution at greater sensitivity.

Table 1. Comparsion of Tenerife and COBE map-making sensitivities

	Tenerife	COBE
Frequency (GHz)	10 15 33	31.5 53 90
Sensitivity (mK $Hz^{-\frac{1}{2}}$)	9 3.3 3.3	28 12 16
Best noise level per beam (μK)	74 29 20	73 28 40
Expected noise after one year (μK)	38 34 34	113 49 64

Table 2. Current sensitivity levels for the Tenerife scans (μK per beam)

	35	37.5	40	42.5	45
10GHz	95	94	59	72	143
15GHz	43	20	30	32	83

3 Production and Inspection of the Initial Maps

To compile the maps calls for the combination of all of the data. The editing, regriding and calibration are the same as previously reported for the declination

+40° scan alone (Davies et al 1987 and Watson et al 1992). The results of this are five final stacked scans at each declination. We have not included the declination +45°scan as it is still very undersampled.

There are two options for the generation of the maps. The first is a simple-minded patching together of the final scans and the second utilises a Maximium Entropy based algorithm (Lasenby el al. in preparation) to discern the underlying sky brightness distribution. Here we have chosen the former method on the grounds that it makes the least assumptions about the data and it is easier to follow the treatment of the noise. The final scans have been smoothed slightly in order to integrate the noise into a beam-size with a small reduction in the resolution. This produces a map which truly reflects the inherent noise level of the data. A simple linear interpolation has been carried out between declinations to produce these composite maps. The noise on each of the declination scans used is different and their average noise levels per beam are given in Table 2. These maps should therefore be regarded as mostly qualitative since the noise varies across the map. A more careful analysis and discussion will be presented in Gutierrez de la Cruz et al (in preparation).

We have used the same greyscale range in both maps and yet the 10GHz map appears to contain more structure. This due to the presence of a mixture of higher instrumental noise and galactic emission. Until the noise level is improved these cannot be separated, although with the steep spectral index of Galactic emission useful upper-limits can still be placed on its contribution to the 15 and 33GHz data. An exception is the detection of a long continuous structure consisting of a broad arc of what most probably is non-thermal galactic emission covering right ascensions 220° to 240°. Two other strong features are point sources 3C345 and 0923+39 and are visible at the predicted amplitudes. 3C345 is a flat spectrum source and is also in the 15GHz data, but 0923+39 has an inverted spectrum which peaks at 10GHz. The 15GHz map is relatively free from structure apart from 3C345 and the identification of a suspect CMB "hot spot" at right ascension 187° and declination +40°. This feature also seems to appear in the 10GHz map at roughly the same amplitude, but is overwhelmed by other features. The feature also appears in the data from the 33GHz instrument which has so far only observed this declination (Hancock et al (submitted)).

4 Preliminary Two Dimensional Statistical Analysis

The method we have used in the past for single scans has been likelihood analysis (Davies et al. 1987). This can be extended to two dimensions by generalising the construction of the covariance matrix so that all possible pairs of points in the map are considered instead of just those within a single scan. This method and detailed results will be reported in Gutierrez de la Cruz et al. (in preparation). The preliminary results from this method using data up to the end of 1992 are very similiar to those for the declination 40° scan alone. They place limits on the 10GHz data of $\delta T/T \approx 3 \times 10^{-5}$, while 15GHz has a 95% confidence upper-limit at $\approx 2 \times 10^{-5}$, with indications of detected structure appearing at the

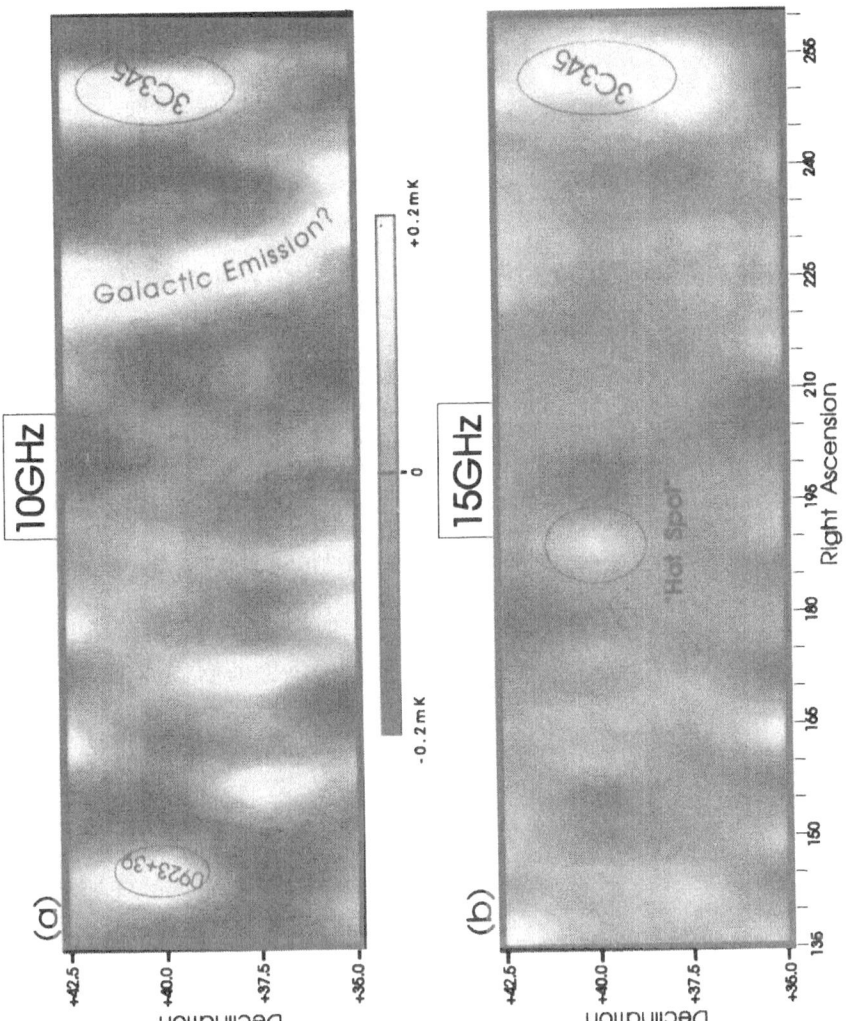

Fig. 1. Tenerife 10 and 15GHz maps of the CMB.

10^{-5} level. All quoted results are for intrinsic perturbations using a Gaussian autocorrelation function with coherence angle $4°$.

5 Conclusions and Future Prospects.

A combination of an improved observing strategy, sensitivity and an excellent spring observing season, has led to notable improvements in the signal to noise on our central declination scans. The Tenerife experiment is therefore well placed to work towards the production of multi-frequency maps, where most of the independent fields contain an appreciable signal. The result of this will be maps in which one can directly see structure at the resolution scale. As yet only one or two declination strips at each frequency have reached the required level to actually resolve structure. This level appears to be at about 20 to 30μK and will require a year or more to produce uniformly sampled maps at this sensitivity. The 33GHz maps should be particularly interesting since the extrapolated galactic emission from the other maps must be less than $\delta T/T \approx 3 \times 10^{-6}$ allowing unambiguous identification of CMB features expected at the 10^{-5} level. A combination of the Tenerife and the COBE maps should allow the determination of the power spectrum from large to intermediate angular scales. Therefore this will provide a useful step between the large scale perturbations found by COBE and the degree scale limits like those from the South Pole (eg Schuster et al, 1993). With the prospect of future degree scale surveys with a new generation of experiments, this sky region should be considered as one of the target areas.

References

R.D. Davies, A.N. Lasenby, R.A. Watson, E.J. Daintree, J. Hopkins, J. Beckman, J. Sanchez-Almeida and R. Rebolo. 1987, Nature, **326**, 462

J. Schuster, T. Gaier, J. Gundersen, P. Meinhold, T. Koch, M. Seiffert, C.A. Wuensche and P. Lubin, 1993, preprint

G.F. Smoot et al, 1992, Ap. J. Lett., **396**, L1

R.A. Watson, C.M. Gutierrez de la Cruz, R.D. Davies, A.N. Lasenby, R.Rebolo, J.E. Beckman and S. Hancock, 1992, Nature, **357**, 660

R.A. Watson and C.M. Gutierrez de la Cruz, 1993, (Accepted Ap.J.Lett.)

ANISOTROPY of the RELIC RADIATION in RELICT-1 EXPERIMENT and PARAMETERS of GRAND UNIFICATION

M. V. Sazhin[1] , I. A. Strukov[2] , A. A. Brukhanov[2] , D. P. Skulachev[2]

[1] Sternberg Astronomical Institute, 119899 Moscow, Russia
    E-mail: snn@sai.msk.su
[2] Space Research Institute, Profsojuznaja, 84/32, Moscow,
    Russia, 117810

1 Introduction

In January 1992 in the Sternberg Astronomical Institute, Moscow at the Astrophysical Seminar it was announced that after additional processing the RELIKT-1 data of the 1983-84 space survey the large scale anisotropy of the relic radiation has been detected. During 1992 the COBE group of observers announced the detection of the large scale anisotropy of the relic radiation too [1 -5].

Let us discuss scientific information which one can obtain from investigation anisotropy of the CMBR and specific goals of this observation. One can expand the large scale anisotropy of the CMBR as sum of spherical harmonics

$$T(\theta) = T_0 + T_1 \cdot cos(\theta) + T_2 \cdot (\frac{1}{3} - cos^2(\theta)) + \dots$$

the first term is monopole component of the CMBR or the CMBR itself. It was discovered by Penzias and Wilson at 1965. The observation of the CMBR allowed us to make choise between the cold and the hot model of our Universe. Now the hot model of our Universe is generally accepted. Next step in investigation of large anisotropy of our Universe was the observation of dipole anisotropy or the observation and determination of the second term of the sum . The value of dipole anisotropy $3mK$ indicates that our Galaxy moves with respect to relic frame of reference. This observation was a progenitor of the observation of peculiar galactic velocities.

Third term is quadrupole anisotropy. It allow us to make some conclusions about the early Universe.

The large scale perturbations which correspond to spherical harmonics with $2 < l < 30$ are one of the most powerful tool of investigation of the early Universe. They are now in the regime of linear growing. Their amplitude is not affected by late stage processes and are determined only by the parameters of the early Universe. The parameters of the early Universe are defined by super high energy physics namely the Grand Unification Theory (GUT). So,the amplitude of large scale perturbations are defined by the parameters of GUT. It was clarified when the theory of inflation was developed and the theory of perturbation in inflation was also developed [6-9].

The data of amplitude of low spherical harmonics define also a part of the initial spectrum perturbation. If the spectrum is Harrison - Zeldovich type the quadrupole component determines also the initial values of large scale structure of the Universe, as far as the H-Z spectrum is determined completely by one parameter (amplitude of the spectrum). In the case of a modified H-Z spectrum the connection between quadrupole component and parameters of large scale structure of the Universe is more complex.

Here we briefly discuss the results of additional data processing obtained from space experiment RELICT-1, compare it with the COBE quadrupole components and others data, and discuss some possible conclusions for the physics of the very early Universe (GUT parameters); and the nature of contributors of dark matter in the Universe is also discussed.

2 RELICT-1 data reprocessing

The RELICT-1 survey was carried out from the satellite board at the frequency range of 37 GHz with an angular resolution of 6°. Details about the experiment's configuration and data preparation were discussed in previous papers [2, 10-12]. New version of the data processing did not include any simplification in the model and as a result it shows the presence of an anisotropy of the microwave background.

We have corrected the initial data by removing the modelled contribution of the Earth's, Moon's and Sun's radiation. All the data in which this contribution was more than 15 μK (in the smoothed data) were excluded of analysis. Also, we excluded the data in which the difference between observed and modelled data was more then 10% (during fast motion of the satellite near the Moon and the Earth). After this correction, the dipole component [13] and the mean outside the Galactic plane were subtracted from the data. We have made an analysis and an estimation of the signal after the additional smoothing the data on the map.

3 RESULTS OF ANALYSIS

The method of estimation of the signal is to compare the measured value (the sum of the noise and the signal) with the amplitude of the apparatus noise, which is measured with high accuracy [1,2]. In order to estimate the amplitude of the signal we modelled the signal which is determined by the H-Z spectrum for primordial perturbation [14]

$$< \Delta T_l^2/T^2 >= \pi \varepsilon_H^2 (2l + 1)/2l(l + 1), \tag{1}$$

where l is the number of spherical harmonics, and ε_H is the amplitude of the metric fluctuation.

The stochastic signal for the spectrum (1) is modelled onto the map. Then, the complete process of observations, and the data reduction is simulated in order to obtain survey transfer function, the mean value of the signal's dispersion and the variance's dispersion.

The measured values (corrected for effects of the apparatus and for the Sun's, Earth's and Moon's contributions) and the parameters of noise are listed in table 1 for different smoothing of experimental data. The data for a smoothing angle of 24° show that the signal is detected with a probability of 97%.

Table 1. Parameters of noise and measured value

Smoothing parameter angular degree	Measured variance on the map $microK^2$	Variance of the noise $microK^2$	Degrees of freedom of noise
6	127^2	124^2	106
12	67^2	56^2	19
24	34.5^2	23.7^2	9

An anomalous low value of temperature (we called it "blamb") is observed in the region which lies inside the ecliptic longitudes $65° \div 340°$ and ecliptic latitude $-20° \div -45°$.

Statistical simulation shows that the probability that this anomaly is a peak of Gaussian noise on the map is about 1%. With a confidence level of 90% we can estimate the mean value of the signal in this region ΔT_b inside the interval

$$-114\,\mu K < \Delta T_b < -27\,\mu K$$

or

$$-4 \cdot 10^{-5} < \Delta T_b/T < -1 \cdot 10^{-5}.$$

We have analyzed the following possible sources of the signal which have a noncosmological origin: apparatus effects, the Moon's, Earth's and Sun's radiation, and the contributions of Galactic sources.

Modelling the complete process of our survey including Moon, Earth and Sun shows that the observed signal could not be explained completely neither by the radiation of the known radio sources nor by the systematic errors of the survey. We expected the signal has the cosmological origin.

We use spectrum (1) to estimate the signal. The upper and lower limits of the mean quadrupole with a confidence level of 90% are:

$$17\,\mu K < \langle \Delta T_2 \rangle < 95\,\mu K.$$

$$6 \cdot 10^{-6} < \langle \Delta T_2/T \rangle < 3.3 \cdot 10^{-5}$$

and the amplitude of fluctuation ε_H in (1) is inside the interval

$$5.2 \cdot 10^{-6} < \varepsilon_H < 2.9 \cdot 10^{-5}.$$

The COBE data show [4]

$$2.9 \cdot 10^{-6} < \varepsilon_H < 5.5 \cdot 10^{-6}.$$

4 COMPARISON WITH THE COBE AND 19.2 GHz DATA

After the COBE group [3,4,5] declared its new results, it became possible to compare our data with the COBE data. Examining the RMS value of the cosmic quadrupole of the COBE data, one can see that the mean value of COBE quadrupole intensity $17 \pm 5\mu K$ (for 68 % C.L) is less than RELICT one. Taking into account the experimental errors both amplitudes of quadrupoles correspond to one another. But, as one can see below there is significant disagreement in the location of quadrupoles. Unfortunately the COBE group has not yet published complete data, so we are forced to compare our map with COBE quadrupoles only and will analyze the COBE data and these disagreements. We will take into consideration only the declared sensitivity of the different radiometers of the COBE and the statistical arguments.

One can compare the RELICT and COBE quadrupole using cross correlation $r(R \cdot C)$

$$r(R \cdot C) = \frac{\sum Q_i(R)Q_i(C)}{\sqrt{\sum Q_i^2(R) \sum Q_i^2(C)}}$$

where $Q_i(R)$ are RELICT quadrupole components, $Q_i(C)$ are COBE quadrupole components and Q_1 is equal to the real quadrupole coefficient multiplyed by 3/4 the numerical coefficient. This direct comparison shows that RELICT and COBE quadrupoles do not correlate. For example,

$$r(R_{37} \cdot C_{53}) = -0.4$$

$$r(R_{37} \cdot C_{31}) = 0.2$$

where the lower index referees to the frequency of a map.

Therefore, there are two choices from this situation. The first is that the COBE quadrupole is correct and there are some systematical errors or unseparated galactic emission in the RELICT data. The second is that the RELICT quadrupole is correct and there are some systematical errors in the COBE data.

If one believes in the COBE results one should explain the discrepancy between the COBE proposed free-free emission [4] and optical data of [15, 16]. The interpretation of the COBE data requires at least 3 times more powerful free-free emission then follows from the Reynolds observations. We shall assume the Reynolds observations valid and analyze the second choices that the COBE data to some extend are affected by errors.

The COBE team shows a different quadrupole $Q_{(A-B)}$ [17]. In these data the cosmic signal should be reduced to zero as far as the transfer functions of two channels (A and B) are equal to one another. On the other hand, the $\chi^2 = 11$ for 31 GHz channel it indicates the presence of some signal with confidence level 95% . Only one explanation of this big value of χ^2 for 31 GHz channel is possible. It is the presence of a residual systematical error in the channel. To prove this claim one should calculate the correlation between the sum of the two channels and difference of its

$$r_{31}(Q_{(A+B)}, Q_{(A-B)}) = -0.48$$

The lower index 31 refers to the frequency. The correlation coefficient is significant and negative. From the other side, the cosmic signal is completely reduced from the difference map $A - B$. The significant correlation indicates that there exists some signal in the B channel. Seems to be that it is a error. If so, the variance of B channel must be more than the variance of A channel. One can use the variation of the sum of two channels and the variation of the difference to calculate the variation of B and A channels separately. One can obtain $\sigma_B^2 = 3 \cdot \sigma_A^2$, so our assumption on the presence of a systematical errors in B channel seems to be valid.

One can also find the correlation coefficients between A and B channels for different frequencies. The results are shown in Table 2. The data in this table is shown as follows. In the first column there is the list of maps including preliminary COBE map [18], the maps of two frequencies (53 and $90GHz$)., RELICT map and the difference map $(A - B)$ for $31GHz$ and the sum of two channels for $31GHz$. In the second column the correlation coefficients between A channel and the left hand maps are shown. In the next column the correlation coefficients between B channel and the first column are shown. The r of $(A\ B)_{31}$ and B is equal to -0.85 instead of value 0.46 of $(A - B)_{31}$ and A. So, it supports again our assumption that B channel is contaminated by systematical errors.

Therefore, one should use only one channel A to analyze cosmological conclusion of COBE experiment. On the other hand, all data which are correlated with B_{31} channel should be rejected from consideration (or should be corrected).

The additional argument is that channel B_{31} is not correlate with RELICT quadrupole instead of A_{31} channel (see row 6). The correlation between A_{31} and RELICT data is 0.75 which is rather significant.

As far as $53GHz$ map is correlated with B_{31} (see row 4 of the table2) it are also contaminated by systematical errors.

It would be noticed that the direct comparison the COBE and RELICT quadrupole components is not correct due to the incomplete sky coverage in both surveys, low signal to noise ratio and different transfer and weighting functions of this two experiments. In this case the spherical harmonics lose its orthogonality and the power of the some harmonics transfers to others, moreover this transformation is different for COBE and RELICT radio maps.

Table 2. The correlation coefficients

Map	Channel $A(31GHz)$	Channel $B(31GHz)$
31 GHz$_{20cut,(A+B)}$	0.55	0.87
31 GHz$_{20cut,(A-B)}$	0.46	-0.85
53 GHz$_{20cut}$	0.22	0.66
53 GHz$_{30cut}$	-0.13	0.52
90 GHz$_{20cut}$	-0.31	0.47
90 GHz$_{30cut}$	-0.43	0.31
RELICT	0.75	-0.21
Channel $A(31GHz)$	1.00	0.07
Channel $B(31GHz)$	0.07	1.00

It would be very interesting to compare COBE and RELICT data with the 19.2 GHz survey [19] . COBE separation the Galactic and cosmic microwave emission [4] one can calculate quadrupole component at 19.2 GHz has to be 100 or 200 μK depends on the Galactic cut 10 or 20° correspondingly. Such value of the signal may be easy detected with the sensitivity declared for the 19.2 GHz survey. If our conclusion of the cosmological origin of RELICT signal is incorrect, the RELICT signal also has to be detected at 19.2 GHz at the same level.

5 COMPARISON with MODELS

Many authors have computed anisotropies of the relic radiation for different cosmological models. One can find the sum of these efforts in [20]. There are two main conclusion for cosmology which are the results of these detection. One is concerned to new physics and reveals the main parameters of interaction on the energy scale of the order of $10^{16}GeV$ and second is concerned to the present Universe and its contributors.

First of all we would like to concentrate on the parameters of Grand Unification. It is well known that the observational restrictions for $\delta T/T$ are the powerful test for modern theories of particle physics(see, for example, [21, 22] . Now there is hope that large scale anisotropy is discovered. Therefore, it is possible to estimate some main parameters of the elementary particle interaction, which determines the interaction at the energy scale near the Plank scale.

Although the standard minimal $SU(5)$ model with a Coleman - Weinberg potential with a large coupling constant is rejected by experimental data there is possibility that extended $SU(5)$ or supersymmetric $SU(5)$ work. We consider below some models which produce acceptable predictions. One is constructed by Shafi and Vilenkin [23] and second is constructed by Pi [24]. The authors add weakly interacting scalar singlet ϕ (it is real ϕ in Shafi and Vilenkin (SV) model and is complex ϕ in model of Pi). This field is coupled with other physical field with small coupling constant. In SV model the vacuum expectation value of ϕ induces $SU(5)$ symmetry breaking. In Pi model the real component of ϕ drives inflation and the imaginary component of ϕ is an axion field.

Potential which drives inflation can be represented in both models in standard form

$$V(\phi) = \frac{\lambda}{4} \cdot (\phi^2 - \phi_0^2)^2$$

Similar potential appears in supersymmetric models [25, 26].

Supersymmetry has some advantages from theoretical point of view. In cited paper of Ellis et al. analyzed consequences of supersymmetric model described by potential

$$V(\phi, T) = \alpha \cdot \phi^4 - \beta\phi^3 + (\gamma + c \cdot T^2)\phi^2 + \delta$$

where α, β, γ, c are parameters of the model, T is temperature of the plasma and δ is vacuum energy. The estimation which will be done below concerns to α in this model. It is necessary to mention that the difference of this potential from the standard form leads to some numerical coefficient of the order of unity.

Therefore, we can estimate the coupling constant in $\lambda\phi^4/4$ potential and it is approximately valid for more complicated models.

One can write [21] that $\varepsilon_H = 7.6 \cdot \sqrt{\lambda}$ and using the RELICT data we obtain the estimation of coupling constant in these models

$$5 \cdot 10^{-13} < \lambda < 1.5 \cdot 10^{-11}$$

The scale of $SU(5)$ symmetry breaking (in SV model for instance) is determined by $M_{Scale} \approx \lambda^{1/4} M_{pl} \approx 10^{16} GeV$. We choose arbitrary renormalization mass of the model to be M_{pl}. Similar estimation appears in other models of extended $SU(5)$ including supersymmetric models.

The second conclusion is connected with the contributors of our Universe (dark matter). There are authors who elaborated hybrid model in which there is dark matter of two types. One is stable and second is unstable type [27,28]. There is no main defect of standard models in it. It is possible to explaine the existence of the first objects at $z = 4 - 5$ (quasars) and the existence of large scale structure (LSS) at $z = 0$. LSS is evolving very fast in the standard models. In the hybrid model the main contributor consists from hot and unstable dark matter. At $10^{16-17} sec$ the hot particles decay and the rate of evolution decreases. Above mentioned data agree fairly well with the model in which the density perturbation in the moment of recombination is $\simeq 4 \cdot 10^{-3}$. The amplitude of

$\Delta T/T$ depends both on $\Delta\rho/\rho$ and on the content of these stable and unstable components. Our data show that the content of stable dark matter is $10\% - 20\%$ of the content of unstable type.

References

[1] Strukov,I.A., Brukhanov,A.A., Skulachev,D.P., Sazhin,M.V., Pis'ma v Astron. Zh. **18** (1992) 387 (Sov.Astron.Lett.,**18**, 153,(1992))

[2] Strukov,I.A., Brukhanov,A.A., Skulachev,D.P., Sazhin,M.V., Mon. Not. R. astr. Soc. **258** (1992) 37P

[3] Smoot, G. et al., Astrophys. J. **396** (1992) L1.

[4] C.L.Bennett, C.L. et al., Astrophys. J., **396** (1992) L7

[5] Wright et all., Astrophys. J. **396** (1992) L13

[6] Guth, A., Phys. Rev. **D23** (1981) 347

[7] Linde A., Phys. Lett. **129B** (1982) 177

[8] Rubakov, V., et all., Phys. Lett. **115B** (1982) 189

[9] Starobinsky A., Phys. Lett., **117B** (1983) 175

[10] Strukov, I.A.& Skulachev, D.P., Sov. Sci. Rev. Astrophys. and Space Phys. **6** ser.E (1987) 147

[11] Klypin, A.A., Sazhin, M.V., Strukov, I.A. & Skulachev, D.P., Sov. Astron. Letters **13** (1987) 104

[12] Strukov, I. A., Skulachev, D. P., Klypin, A. A. & Sazhin M.V. Large Scale Structures of the Universe. (1987) p.27, eds Audouze et al., IAU.

[13] Strukov, I.A., Skulachev, D.P., Boyarskii, M.N.& Tkachev, A.N. Sov. Astron. Letters **13** (1987) 65

[14] Abbott,L.F. & Wise,M.B., 1984. Astrophys. J. **282** (1984) L47

[15] Reynolds, R.J., Astrophys. J. **282** (1984) 191

[16] Reynolds, R.J., 1990. Proc. IAU Symp. No 139. The Galactic and Extragalactic Background Radiation, p. 157, eds. S. Bowyer & Ch.Lennert (Dordrecht: Kluwer Academic Publisher).

[17] Kogut, A., et al. Astrophys. J. **401** (1992) 1

[18] Smoot, G., et al., Astrophys. J. **371** (1991) L1

[19] Boughn, S.P., Cheng, E.S., Cottingham, D.A. and Fixsen, D.J. Rev. Sci. Instr. **61** (1990) 158

[20] Inflationary cosmology (1986),Ed.Abbott,L.F. and So-Young Pi, World Sci.

[21] Linde, A., Particle Physics and Inflationary (1990) Cosmology, Harwood.

[22] Dolgov, A.D., Sazhin, M.V. & Zeldovich, Ya.B., Basic of Modern Cosmology (1990) Editions Frontiers.

[23] Shafi Q., Vilenkin A., Phys. Rev. Lett. **52** (1984) 691

[24] Pi, S-Y., Phys. Rev. Lett. **52** (1984) 1725

[25] Ellis J., Nanopoulos D.V., Olive K.A., Tamvakis K., Nuclear Phys. **B221** (1983) 524

[26] Holman R., Ramond P., Ross G.G., Phys. Lett. **137B** (1984) 343

[27] Berezhiani Z.G., Khlopov M.Yu., Zeitschrift fur physik , ser.C.,**49** (1991) 73

[28] Doroshkevich A.G., Pis'ma v Astron.Zhurn. **14**, (1988) 296

RELIKT1 and COBE-DMR* Results: A Comparison

A.J. Banday

Universities Space Research Assoc., Code 610.3, NASA/GSFC, Greenbelt MD, 20771.

Abstract: A recent reanalysis of the RELIKT1 data (Strukov et al., 1992) shows a statistically significant decrement in temperature in a highly smoothed sky-map. COBE-DMR has observed the same region of sky at higher signal-to-noise ratio at three discrete wavelengths. We present results from the DMR sky-maps smoothed to equivalent resolution as RELIKT1 and discuss possible sources of contamination.

1 Introduction

RELIKT1 and the COBE-DMR are to date the only two dedicated space-borne experiments to attempt to detect and map the anisotropy of the Cosmic Microwave Background (CMB). A recent reanalysis of the RELIKT1 data (Strukov et al., 1992) has detected a signal reputedly of cosmological significance at -20° $< \lambda < 65°$, -45° $< \beta < $ -20° (the so-called 'blamb' region). The COBE-DMR has detected a statistically significant signal on the sky which cannot be attributed to known systematic instrumental errors or Galactic emission (see Smoot et al., 1992; Bennett et al, 1992; Kogut et al, 1992; Wright et al, 1992). This structure is consistent with a description in terms of scale-invariant fluctuations with a Gaussian distribution of amplitudes and random phases. However, COBE-DMR has only claimed a statistical detection of anisotropy, and no claim has been made about individual features on the maps. The unambiguous identification of anisotropy would be useful

- As an aid to searches for small-scale anisotropy.
- As a probe of cosmological models from the properties of hot spots and cold spots (Bond & Efstathiou, 1987).

We therefore reanalyse the DMR skymaps using the same technique adopted by the RELIKT1 group.

---

* The National Aeronautics and Space Administration/Goddard Space Flight Center (NASA/GSFC) is responsible for the design, development, and operation of the Cosmic Background Explorer (COBE).

2 Analysis Technique

A weighted sum of the 2 independent A and B channels at each frequency was formed to increase sensitivity, and then the 3 resulting 'sum' maps were masked to reproduce the RELIKT1 sky coverage. The analysis technique from Strukov et al. was then reproduced on the DMR maps, that is

- the best fit monopole and dipole were removed from each map, assigning zero weight to pixels within ± 15° of the Galactic plane
- the maps were smoothed with a Gaussian beam chosen to match the final smoothings of the RELIKT1 map, (FWHM ~ 15°, 29°, 57°), assigning zero weight to pixels within ± 15° of the Galactic plane
- the mean temperature in the 'blamb' region was evaluated.

3 Discussion

Figure 1 shows the RELIKT1 and DMR maps in an ecliptic projection. It is clear from the figures that

- the DMR maps have lower noise than the RELIKT1 map
- the only obvious features in the DMR maps have spectral indices suggestive of a Galactic origin (corresponding to structure associated with Ophiuchus and Orion lying outside the Galactic exclusion zone).
- the DMR maps show no evidence for the blamb.

Table 1 gives the weighted average temperatures in the blamb region. The RELIKT numbers are from Strukov et al. (1992). The errors are 68% c.l. and include systematic error estimates. The DMR results are consistent with no statistically significant structure in the blamb region. Table 2 gives the likely contribution from Galactic foregrounds, determined at the three DMR frequencies from the models described in Bennett et al. (1992). The RELIKT1 numbers are again from Strukov et al. Our conclusion remains unchanged if we correct for the Galactic foreground signals.

	T (μK)
RELIKT1	-71 ± 27
DMR 31 GHz	-4 ± 11
DMR 53 GHz	-1 ± 4
DMR 90 GHz	+5 ± 5

Table 1. Sky temperatures in the direction of the claimed RELIKT1 fluctuation.

Term	T (μK)			
	31 GHz	53 GHz	90 GHz	RELIKT1
Synchrotron	-1	0	0	< 13
Free-Free	-12	-4	-1	< 3
Dust	0	-1	-2	< 1

Table 2. Estimate of the Galactic contribution.

4 Conclusions

We find no evidence for the temperature decrement claimed by the RELIKT1
group. It is unlikely that systematic errors or Galactic foregrounds have affected
our analysis.

References

Bennett, C.L. et al., 1992, *Astrophys. J. Letts.*, **396**, L7.

Bond, J.R. & Efstathiou, G., 1987, *Mon. Not. R. astr. Soc.*, **226**, 655.

Kogut, A. et al., 1992, *Astrophys. J.*, **401**, 1.

Smoot, G.F., et al. 1992, *Astrophys. J. Letts.*, **396**, L1.

Strukov, I.A. et al., 1992, *Mon. Not. R. astr. Soc.*, **258**, 37P.

Wright, E.L. et al., 1992, *Astrophys. J. Letts.*, **396**, L13.

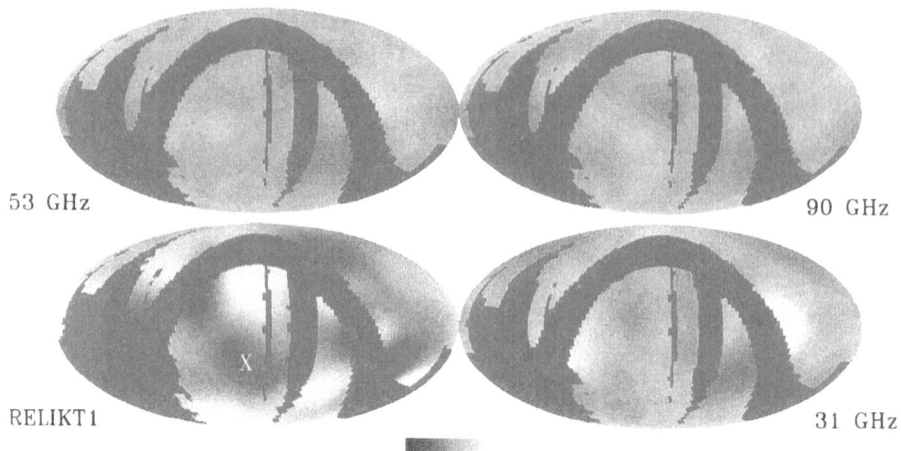

Figure 1: RELIKT1 and DMR sky maps in ecliptic coordinates smoothed to an effective FWHM of $\sim 29°$. The temperature scale is thermodynamic. A band of width $\pm 15°$ about the Galactic plane has been given zero weight, other blank regions correspond to the RELIKT1 sky coverage. The white cross on the RELIKT1 map shows the approximate centre of the 'blamb' region.

Comments on the *COBE*[1] DMR Quadrupole Estimation

L. Tenorio[2], *G.F. Smoot*[2], *C. Lineweaver*[2], *G. Hinshaw*[3], *A. Banday*[4]

[1] The National Aeronautics and Space Administration/Goddard Space Flight Center (NASA/GSFC) is responsible for the design, development, and operation of the Cosmic Background Explorer (*COBE*). Scientific guidance is provided by the *COBE* Science Working Group. GSFC is also responsible for the development of the analysis software and for the production of the mission data sets.
[2] LBL, SSL & CfPA, Bldg 5-351, University of California, Berkeley CA 94720, USA
[3] Hughes STX Corporation, Mail Code 685.9, NASA Goddard Space Flight Center, Greenbelt, MD 20771 USA
[4] Universities Space Research Association, Mail Code 685.9, NASA Goddard Space Flight Center, Greenbelt, MD 20771 USA

1 Introduction

The *COBE* DMR team ([Bt],[Sm1],[Sm2], [Wt1],[Wt2]) has reported the detection of large angular scale anisotropy in the cosmic microwave background. One measure of this anisotropy is the reported rms amplitude, Q_{rms}, of the quadrupole moment of the observed sky temperature fluctuations. The components of the quadrupole are determined through a least-squares fit of the data to spherical harmonics. Since Q_{rms} is based on a quadrature sum of the measured components, all power including noise power contributes positively to it. To be more precise, suppose the five measured quadrupole components are $Q_i = q_i + \epsilon_i$, where $\{q_i\}$ are the real quadrupole components and $\{\epsilon_i\}$ are the measurement errors with zero mean and variance $\mathrm{var}(\epsilon_i) = \sigma_i{}^2$. The reported Q_{rms} is $\mathcal{Q}(Q_i)^{1/2}$, where $\mathcal{Q}(Q_i) = (\Sigma_i Q_i^2)/4\pi$. It is an estimate of the real amplitude $\mathcal{Q}(q_i)^{1/2}$. An unbiased estimate of $\mathcal{Q}(q_i)$ is

$$\mathcal{Q}(q_i) \approx \mathcal{Q}(Q_i) - \mathcal{Q}(\sigma_i). \qquad (1)$$

Gould [Gd] points out that if the σ_i's are significant compared to the Q_i's, the reported Q_{rms} should be corrected for the noise quadrupole power $\mathcal{Q}(\sigma_i)$. He uses the published σ_i's and equation (1) to determine a noise corrected Q_{rms}. In addition, Stark [St] explains that one must also account for the effects of aliasing of higher order multipoles due to Galactic masking, non-uniform sky coverage and pixelization. He determines σ_i's due to these effects and uses (1) to correct Q_{rms}.

   We present results of Monte Carlo simulations which quantify the effects that pixelization, non-uniform sky coverage, Galactic masking and cosmic variance have on the quadrupole estimates Q_{rms}. We obtain corrections for these effects which are 10% and 25% smaller than those reported by Stark and Gould.

2 Simulation Techniques

The DMR instrument measures differential temperatures of the sky at frequencies 31, 53 and 90 GHz. At each frequency measurements are made with two nearly independent channels A and B. After the data have been calibrated and corrected for known effects (see [Sm1] and [Kg]), an inversion process leads to temperature skymaps at each frequency and channel. At each frequency we form the sum and difference maps by adding and subtracting the two channels pixel by pixel. That is, $T_{sum}(i) = (T_A(i) + T_B(i))/2$ and $T_{dif}(i) = (T_A(i) - T_B(i))/2$ for each pixel i. The sum maps preserve the sky temperature while the difference maps are expected to consist purely of noise

$$T_{sum}(i) = T_{sky}(i) + n_i$$
$$T_{dif}(i) = n'_i. \tag{2}$$

The noise n_i, n'_i in the maps is Gaussian with zero mean and variance that depends on the number of times $N_{obs}(i)$ each pixel was observed

$$var(n_i) = var(n'_i) = \sigma_\nu^2/N_{obs}(i). \tag{3}$$

Here σ_ν is a known instrument rms at a given frequency.

To simulate the DMR skymaps (2) we Monte Carlo skies with a scale invariant power spectrum of fluctuations with a quadrupole-normalized amplitude of $16\mu K$: the coefficients $\{b_{\ell m}\}$[5] of the spherical harmonic expansion $\Sigma b_{\ell m} F_{\ell m}$ are chosen to be Gaussian with zero mean and variance given by

$$var(b_{\ell m}) = < Q_{rms}^2 > \frac{24\pi}{5\ell(\ell+1)}, \tag{4}$$

with normalization $< Q_{rms}^2 >^{0.5} = 16\mu K$. Due to cosmic variance, the individual simulated skies do not necessarily possess a $16\mu K$ quadrupole or a pure $n = 1$ spectrum of higher order power. In some of our simulations we suppress cosmic variance by normalizing the $b_{\ell m}$ coefficients for each order of ℓ to produce a pure $n = 1$ spectrum and quadrupole amplitude of $16\mu K$. Noise is generated to simulate the one-year 53 GHz A-B map; i.e., each pixel temperature T_i is a Gaussian with zero mean and variance (3) with $\sigma_\nu = 18.2mK$. All temperatures referred to in this paper are thermodynamic.

To find the quadrupole components we do a weighted least-squares fit to minimize the contribution from the noisy pixels; we minimize the weighted sum of squares

$$\chi^2 = \sum_i [T(i) - \sum_{\ell=0}^{2} \sum_{m=-\ell}^{\ell} b_{\ell m} F_{\ell m}(i)]^2 / \sigma(i)^2$$

with respect to the $b_{\ell m}$'s, the sum runs over all the pixels outside the Galactic cut and $F_{\ell m}(i)$ are the spherical harmonics evaluated at pixel i. Once we have the fitted b_{2m}'s we compute the Q_{rms} using $Q_{rms}^2 = Q(b_{2i})$.

---

[5] We assume the normalization in which the $\{b_{2m}\}$ are the quadrupole coefficients $\{q_i\}$.

3 Simulation Results

Figures 1-3 show how the different effects accumulate to produce a bias. The median of the input simulations is drawn as a dotted line on the plots. The effects of pixelization and Galactic masking are shown in Fig.1. Noise and non-uniform sampling are added for Fig.2, and finally, cosmic variance is added to the simulations of Fig.3. In all the plots the x axis is the range in Galactic latitude $|b|$ that has been removed from the maps before the fit was done. The y axis is the resulting Q_{rms} in μK of the fit. The upper set of curves include power in the multipole range $2 \leq \ell \leq 20$. The lower set of curves include no intrinsic power in the quadrupole moment, but includes power in the multipole range $3 \leq \ell \leq 20$. While the latter is not expected to represent a plausible physical model of the CMB anisotropies, it is included to aid in assessing the magnitude of the quadrupole bias due to aliasing effects. The five curves in each set correspond to the median, 68% and 95% confidence levels. In Fig.4 we show only the bias of non-uniform sampling, instrument noise and Galactic cut, no multipole structure has been added to the simulations. All curves are derived from 1000 simulated maps.

Figure 1 shows the aliasing effect of the Galactic cut in the presence of higher order power. The Galactic cut aliases power of $\ell > 2$ into the quadrupole, mostly from the multipole $\ell = 4$. Notice that for no Galactic cut the curves converge to the input values, so pixelization is not an important factor. This is to be expected since the 2.6° pixels have an effective $\ell \gg 2$. Figure 2 includes the effect of non-uniform sampling and instrument noise. It shows how the noise pattern and Galactic masking affect the DMR measurement of the Q_{rms} under the assumption that the temperature sky in our horizon is a perfect $n = 1$, $< Q_{rms}^2 >^{0.5} = 16\mu$K sky. In Fig.3 we have added the effect of cosmic variance. In this case, the values of Q_{rms}^2 are distributed like a χ^2 with 5 degrees of freedom. As a result the median value of Q_{rms}^2 is 87% of its mean value which corresponds to 222.6μK^2. In Figure 4 we have simulated only instrument noise to demonstrate the magnitude of the noise bias alone. Note that because the instrument noise is uncorrelated with the higher-order power on the sky, the biases in Q_{rms}^2 due to noise alone and aliasing alone simply add.

The results of Table 1 give the bias power, $Q(\sigma_i)$, for $|b| > 10°$ and 20° of the effects considered. The numbers are obtained by taking the difference of the median of the simulated values and the median of the input values.

4 Conclusions

Equation (1) and the numbers in table 1 can be used to correct the published values of Q_{rms}. For non-uniform sampling, instrument noise and Galactic cut the corrections correspond to the sum in quadrature of the Gould and Stark adjustment (Table 4 in [St]). Our corrections $Q(\sigma_i)$ are about 25% smaller than those of Stark. The corrections due only to noise are about 10% smaller than those in [Gd].

Table 1. Bias of the best-fit Q^2_{rms} for $|b| > 10°$ and $20°$

Figure	Effect	Bias, $|b| > 10°$ μK^2	Bias, $|b| > 20°$ μK^2
1	Galactic cut	7.1	39.8
2	+ non uniform noise	24.7	63.1
3	+ cosmic variance	31.7	65.
4	noise only no structure	17.3	26.2

Acknowledgements: We acknowledge the excellent work of the DMR team as well as the staff and management of the *COBE* project. We thank L. Cayón, E. Martinez and J. Sanz for having given us the opportunity to participate in a wonderful workshop. We thank the staff and management of Palacio de la Magdalena whose smiles got us moving every morning in that Kingdom by the Sea. We are greatful to A. Kogut and P. Stark for numerous conversations. We also acknowledge the support of The Office of Space Sciences and Applications of NASA Headquarters.

References

[Bt]   Bennett, C., *et al.* *Ap. J.* **396** 1992, L7
[Gd]   Gould,A., *Ap. J.* **403** 1993, L51
[Kg]   Kogut, A., et al. *Ap. J.* **401** 1992, 1
[Sm1] Smoot, G.F.,*et al.* *Ap. J.* **371** 1991, L1
[Sm2] Smoot, G.F., *et al.* *Ap. J.* **396** 1992, L1
[St]   Stark,P.B., *Ap. J.* **408** 1993, L73
[Wt1] Wright, E.L., *et al.* , *Ap. J.* **396** 1992, L13
[Wt2] Wright, E.L., *et al.* , *Ap. J.* **421** 1994

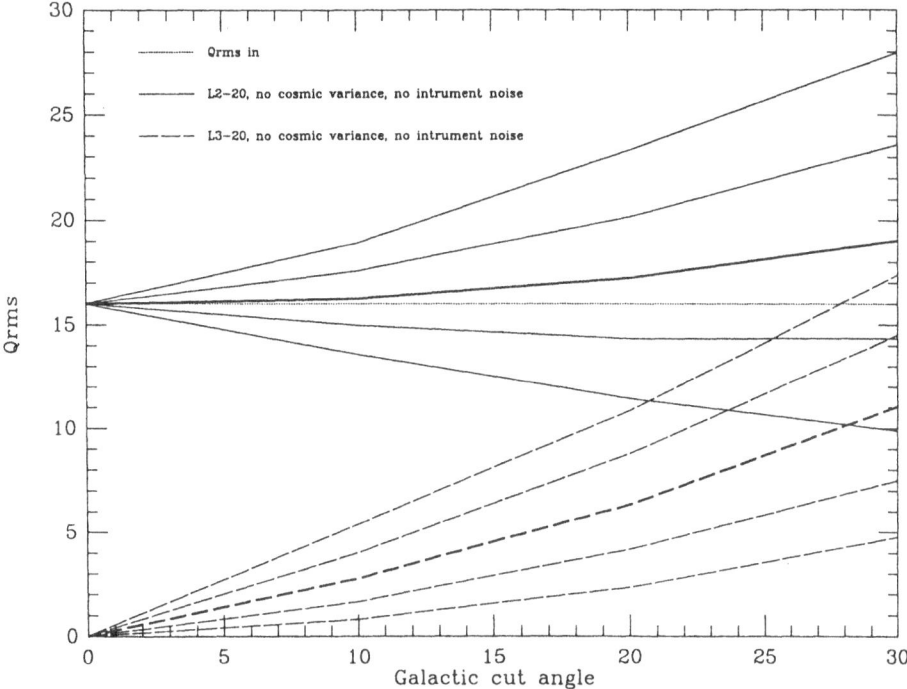

Fig. 1. Effect of Galactic masking

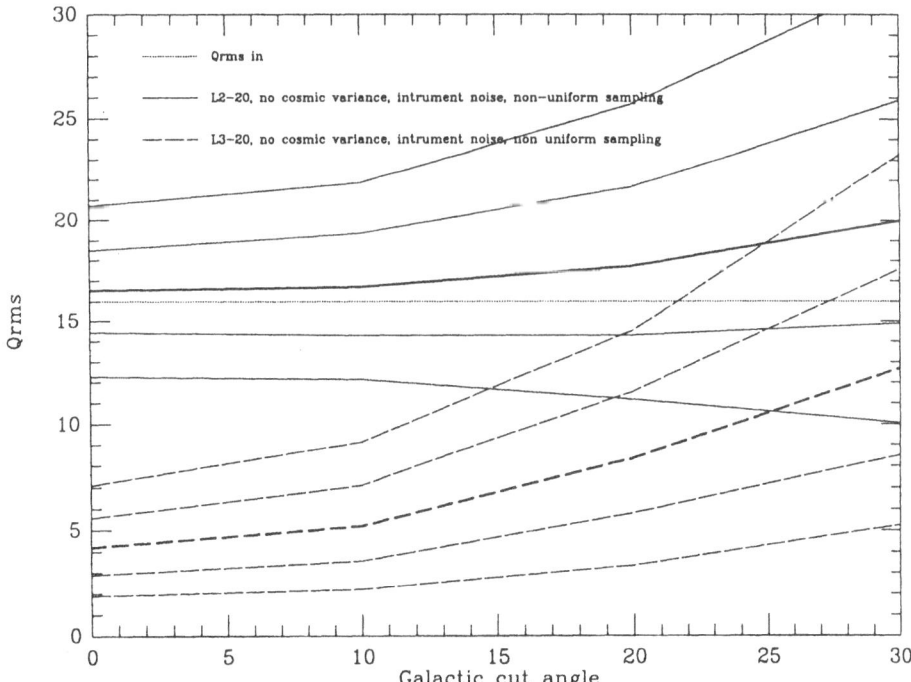

Fig. 2. Effects of Galactic masking, non-uniform sampling and instrument noise

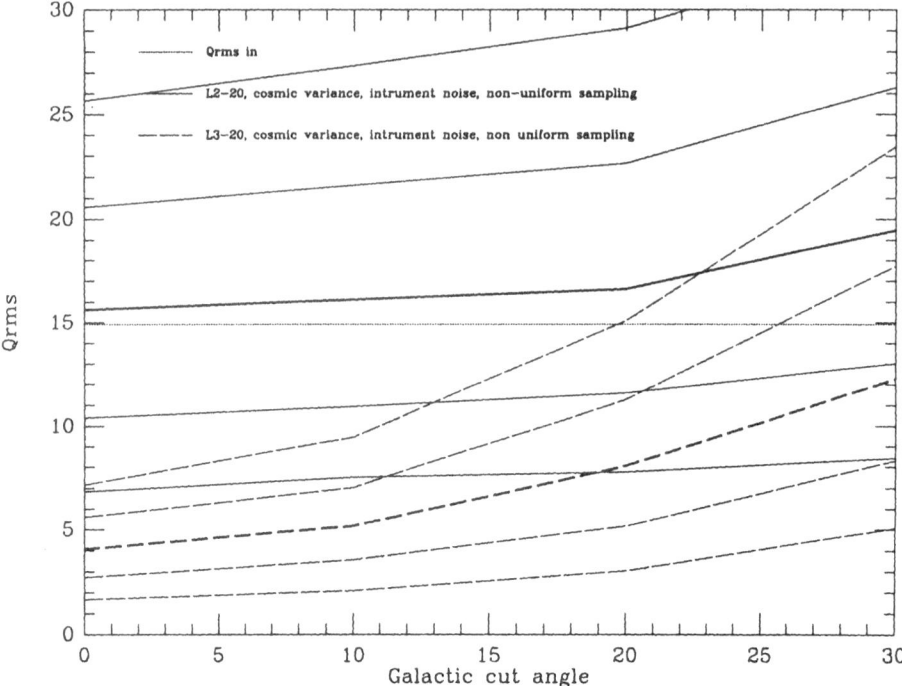

Fig. 3. Effects of Galactic masking, non-uniform sampling, instrument noise and cosmic variance

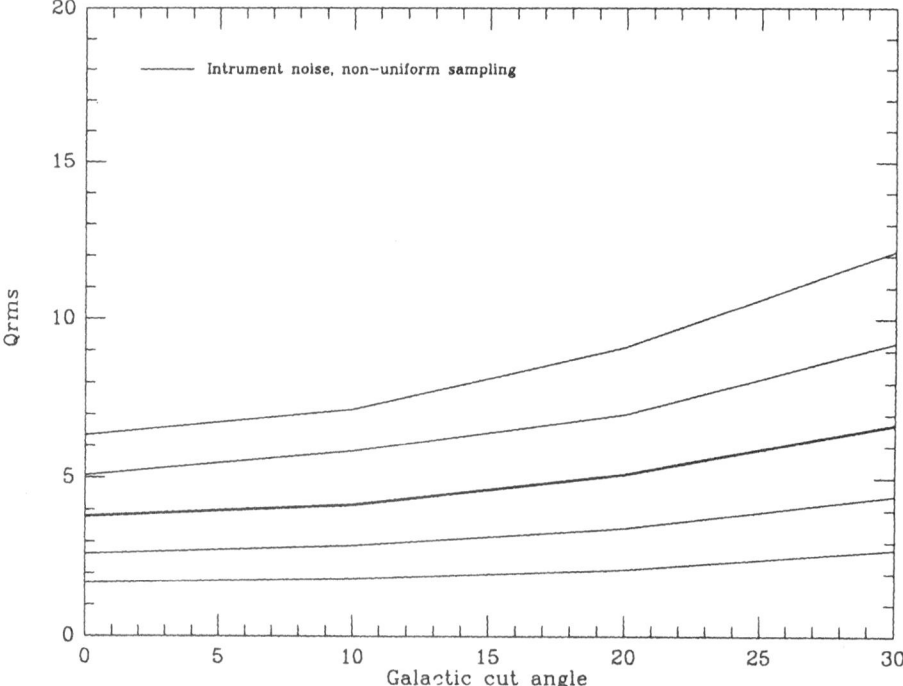

Fig. 4. Bias due to intrument noise, non-uniform sampling and Galactic cut. No structure has been added

Pip Analysis of the Tenerife and ULISSE Data

Laura Cayón[1], Enrique Martínez-González[1], Carlos Gutiérrez de la Cruz[2] and Jose L. Sanz[1]

[1] Dpto. de Física Moderna, Facultad de Ciencias, Universidad de Cantabria. Avda. Los Castros s/n 39005 Santander, Spain.
[2] Instituto de Astrofísica de Canarias. 38200 La Laguna, Tenerife, Spain

1 Introduction

Anisotropies of the Cosmic Microwave Background (CMB) temperature fluctuations at large angular scales (greater than the horizon scale at decoupling time $\sim (2\Omega^{1/2})^{\circ}$) contain information of the spectrum of matter density fluctuations at scales above hundred Mpcs. There are several experiments working at these scales. The level of anisotropy was first detected by the DMR experiment on board of the COBE satellite giving an amplitude of 1.1×10^{-5} on angular scales of $\sim 10°$ [1]. A cross-correlation analysis between the DMR data and the results of the MIT experiment [2] reveal that the fluctuations observed by both experiments are consistent with CMB anisotropies [3]. The Tenerife [4] and the ULISSE [5] experiments gave upper limits of $\Delta T/T < 1.8 \times 10^{-5}$ and $\Delta T/T < 1.3 \times 10^{-5}$ respectively at angular scales of $\sim 6°$. The former one is presently working at three different frecuencies taking measurements at several declinations and it could reach the level of detection in a short period of time.

The non ergodic behaviour of the CMB temperature fluctuation field is not taken in account in the standard method to analyse the CMB data. This method is based on a bayessian aproach and uses the likelihood function to determine upper limits of the temperature fluctuations assuming that this field is gaussian distributed over the region sampled by the experiments. However, primordial gaussian fluctuations may appear non gaussian distributed over the observed region due to the non ergodic behaviour of the temperature field. Therefore we propose alternative methods to analyse the CMB data. These methods are based on geometrical properties and we have applied them to the one dimensional cases of the Tenerife and ULISSE experiments.

This contribution is organized as follows. In the next section we discuss the ergodicity of the CMB temperature fluctuation field. Section 3 presents the "Pip analysis" and its results for the Tenerife and ULISSE experimets. The main conclusions are presented in section 4.

2 Statistics of the CMB temperature fluctuations

The non ergodic behaviour of the CMB temperature fluctuation field can be seen through the cosmic uncertainty associated to the spherical harmonic moments [6] and to the temperature autocorrelation function [7] and through the non null dispersion of the third and fourth order moments of the temperature field [8]. In relation to the last property one can assume primordial gaussian temperature fluctuations and can calculate the dispersion of the third and fourth order moments over the whole celestial sphere, defined by:

$$\sigma_n \equiv [<(C^{(n)})^2> - <C^{(n)}>^2]^{1/2}, \tag{1}$$

where the averages are taken over the statistical ensemble and
$C^{(n)} \equiv <(\Delta T/T)^n>_{sky}, n = 3, 4$.

We have calculated those quantities for four different experimental configurations and the results are presented in table 1: Two one-beam experiments: DMR ($\sigma = \sqrt{2} \times 3.2\ deg$) and an experiment with the Tenerife beam width ($\sigma = 2.4\ deg$). A double beam switching experiment: Tenerife ($\sigma = 2.4\ deg$, $\alpha = 8.1\ deg$). A single beam switching experiment: ULISSE ($\sigma = 2.2\ deg$, $\alpha = 6\ deg$).

Table 1. Percentage of dispersion of the third and fourth order moments for different experimental configurations.

	COBE-DMR	Tenerife-1	Tenerife	ULISSE
$\sigma_3 / <C^{(2)}>^{3/2}$	22	16	6	7
$\sigma_4 / <C^{(2)}>^2$	94	69	34	39

The non null values of the dispersion of the third and fourth order moments are due to the non ergodic behaviour of the CMB temperature fluctuations. As one can see the dispersion of both moments grow with the antenna beam width and their amplitudes are smaller for beam switching configurations than for one beam experiments (beam switching experiments filter the low order multipoles which are the less ergodic ones).

It is also important to consider the region of the celestial sphere observed by the experiments. Figure 1 shows the bands sampled by the Tenerife and the ULISSE experiments. We made 5000 simulations of the CMB fluctuations as seen by both experiments, based on the spherical harmonic expansion:

$$\Delta \equiv \Delta T(\mathbf{n}) = \sum_{l=2}^{\infty} \sum_{m=-l}^{l} a_l^m Y_l^m(\mathbf{n}), \tag{2}$$

where $\mathbf{n}$ is the unit vector in the direction of observation and the coefficients a_l^m are assumed to be gaussian distributed with $<a_l^m a_k^n> = 0$ unless $k = l$ and

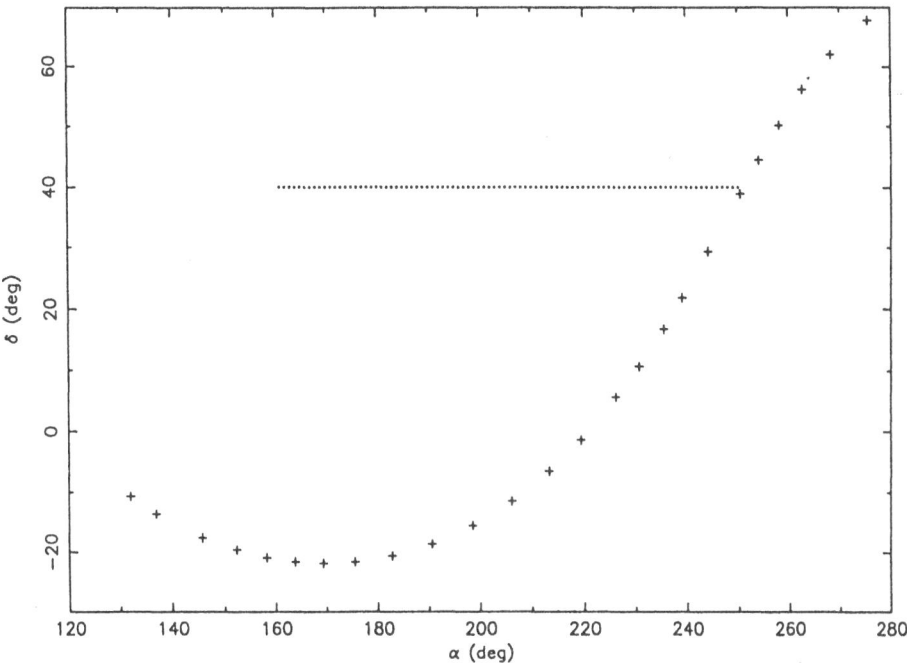

Fig. 1. Region of the celestial sphere observed by the Tenerife (dots) and ULISSE (crosses) experiments.

$m = n$. We also assumed a Harrison-Zeldovich power spectrum $P(k) \propto k$. Table 2 presents the values of the dispersion of the third and fourth order moments obtained for the Tenerife and ULISSE experimental configurations.

Table 2. Percentage of dispersion of the third and fourth order moments for Tenerife and ULISSE experimental configurations considering the observed band of the sky.

	Tenerife	ULISSE
$\sigma_3 / < C^{(2)} >^{3/2}$	52	49
$\sigma_4 / < C^{(2)} >^2$	323	202

As can be noticed comparing table 1 and table 2, as the sampled area decreases the non ergodic behaviour increases producing higher values of σ_3 and σ_4. The bigger region observed by the ULISSE experiment favours the ergodicity of the CMB temperature fluctuation field.

Summarizing, the non ergodic behaviour of the temperature field causes cosmic uncertainties on the spherical harmonics and on the autocorrelation function and produces non null values of the dispersion of the third and fourth order moments. Beam switching tecniques favour the ergodicity although the small area of the sky sampled by an experiment can increase this property. This behaviour should be considered in the CMB data analysis and in the next section we propose geometrical methods to determine upper limits on the CMB temperature fluctuations.

3 Pip Analysis: application to the Tenerife and ULISSE data

We consider a geometrical analysis of the one dimensional data of the Tenerife [4] and ULISSE [5] experiments. The main property of the geometrical methods is the non assumption of ergodicity. In one dimension the up and down crossings of a certain threshold are called "pips" and are characterized by their size and the number of them above different thresholds [9]. The spot is the pip counterpart in two dimensions and one can use the number and area of spots [10, 11, 12, 13, 14], the fractional area of excursion regions, the contour length and the genus [15, 16] to analyse two dimensional data.

The Pip analysis is based on simulations of the CMB temperature fluctuations on the observed band of the sky using a spherical harmonic expansion as indicated in the previous section. We assumed an original gaussian temperature field and a Harrison-Zeldovich power spectrum. The experimental configuration is taken in account through the dispersion of the signal. In the case of a double beam switching experiment, as the Tenerife one, the dispersion of the signal is given by:

$$\sigma_s = (1.5C(0,\sigma) - 2C(\alpha,\sigma) + 0.5C(2\alpha,\sigma))^{1/2}, \tag{3}$$

and in a beam switching configuration, as the one used by the ULISSE experiment, the dispersion of the signal has the following expression:

$$\sigma_s = (2[C(0,\sigma) - C(\alpha,\sigma)])^{1/2}, \tag{4}$$

where $C(\alpha,\sigma)$ is the temperature autocorrelation function, α is the angular separation and σ is the antenna beam width.

Two methods are used to analyse the Tenerife and ULISSE data based on the distribution of the pip sizes and of the number of pips above several thresholds.

One can determine an upper limit of the temperature fluctuations studying the distribution of the pip sizes above the zero threshold. First of all one has to determine the probability distribution function of the pip sizes considering the experimental noise σ_n and the dispersion of the signal σ_s. For this purpose we made 5000 simulations for each value of σ_s in the range $0 - 100 \ \mu K$ for the experimental configurations of the Tenerife and ULISSE experiments. The likelihood function of a set of pip sizes $\theta_1, ..., \theta_N$ for a certain signal $L(\theta_i/\sigma_s)$ is proportional to the product of the probability distribution functions $P(\theta_i/\sigma_s, \sigma_n)$.

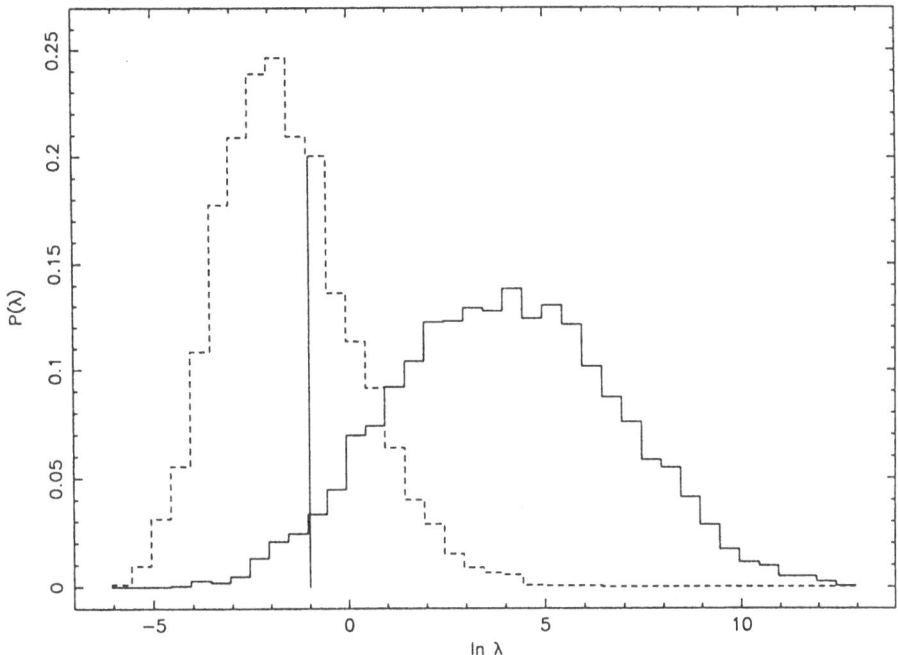

Fig. 2. Distribution of λ for $\sigma_s = 65\ \mu K$ (solid line) and for null signal (dashed line). The vertical line indicates the value of λ_{obs} for the Tenerife data. From this distributions $1 - \alpha = 95\%$ and $\beta = 71\%$.

We used the likelihood ratio as an statistic to determine the upper limit of the temperature fluctuations:

$$\lambda \equiv \frac{L(\theta_i/\sigma_s = \sigma_o)}{L(0_i/\sigma_s - 0)}. \tag{5}$$

Two hypotheses are assumed, H, the cosmological signal is equal to a certain value σ_o and K, the cosmological signal has a null value. The probability of rejecting H when it is true is given by a parameter α and the power of the test (the probability of rejecting K when it is true) is given by β:

$$\alpha = P(\lambda < \lambda_{obs}/\sigma_s = \sigma_o), \beta = P(\lambda < \lambda_{obs}/\sigma_s = 0), \tag{6}$$

where λ_{obs} is the likelihood ratio for the observed pip sizes. Performing 5000 simulations with $\sigma_s = \sigma_o$ and 5000 with $\sigma_s = 0$ one can obtain the probability distribution function of the likelihood ratio in both cases. The parameters α and β are determined comparing with the λ_{obs} obtained for the data.

We obtained an upper limit of 65 μK for the Tenerife data (see figure 2) at the 95% confidence level ($\alpha = 0.05$) with a power of 71% ($\beta = 0.71$). In the case

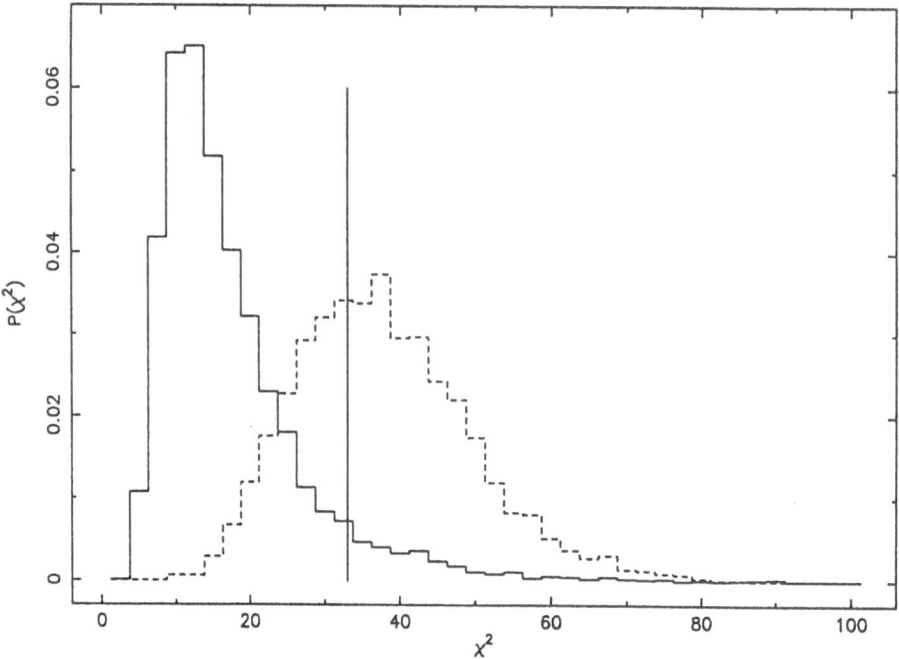

Fig. 3. Distribution of χ^2 for $\sigma_s = 90 \, \mu K$ (solid line) and for null signal (dashed line). The vertical line indicates the value of χ^2_{obs} for the Tenerife data. From this distributions $1 - \alpha = 95\%$ and $\beta = 58\%$.

of the ULISSE data the information about the pip sizes is lost due to the low angular resolution and no upper limit could be determined.

The second method we applied to analyse the Tenerife and the ULISSE data is based on the number of pips above several thresholds. First, one has to calculate the mean number and the dispersion of the number of pips above different thresholds assuming values of the dispersion of the signal in the range $0 - 100 \, \mu K$. 5000 simulations were performed for each value of the signal.

The statistic we used to determine the upper limit is defined by:

$$\chi^2 \equiv \sum_{i=1}^{n} \frac{(N_i - \bar{N}_i)^2}{\sigma_i^2}, \tag{7}$$

where n is the number of thresholds, N_i is the number of pips above the threshold i, $\bar{N}_i$ is the mean number of pips and σ_i is the dispersion of the number of pips. Assuming two hypotheses, in the sense indicated above, the confidence level is given by $\alpha = P(\chi^2 > \chi^2_{obs}/\sigma_s = \sigma_o)$, beeing the power of the test $\beta = P(\chi^2 > \chi^2_{obs}/\sigma_s = 0)$.

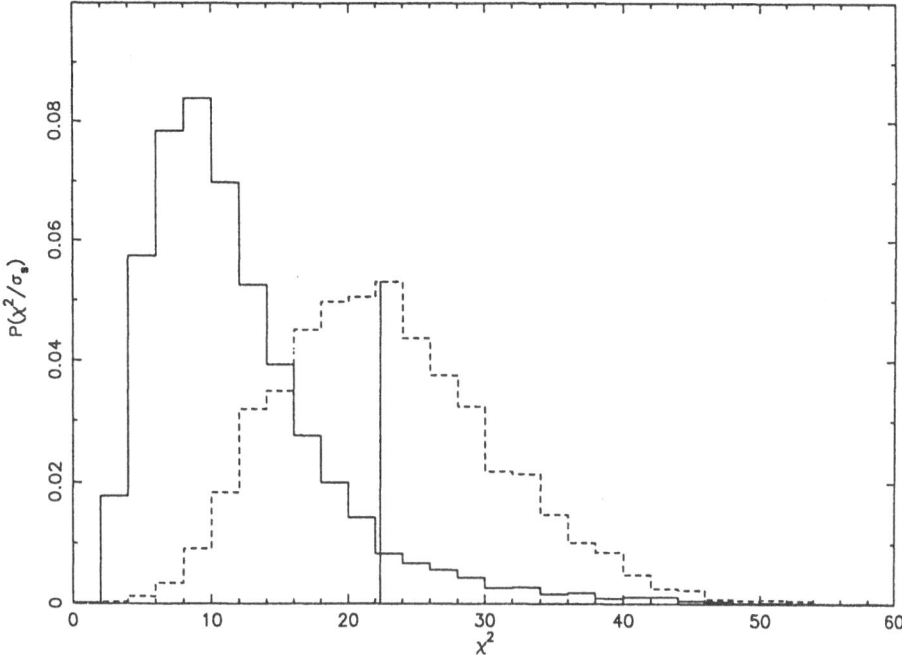

Fig. 4. Distribution of χ^2 for $\sigma_s = 40\ \mu K$ (solid line) and for null signal (dashed line). The vertical line indicates the value of χ^2_{obs} for the ULISSE data. From this distributions $1 - \alpha = 95\%$ and $\beta = 59\%$.

Applying that test we obtained an upper limit of 90 μK for the Tenerife data at the 95% confidence level and power of 58% and an upper limit of 40 μK for the ULISSE data at the same confidence level and with power of 59% (see figures 3 and 4).

4 Conclusions

The non ergodic behaviour of the CMB temperature fluctuations has been seen through the non null dispersion of the third and fourth order moments of this field. Beam switching tecniques favour the ergodicity although the small area covered by the experiments unfavour it.

Geometrical methods have been proposed to consider the no ergodicity of the CMB temperature fluctuation field. An important characteristic of these methods is that one can assume any distribution of the primordial temperature field. In case of one dimensional data they are based on the size of the pips and on the number of them above several thresholds. We applied both methods to analyse the Tenerife and the ULISSE data. The former one is not sensitive to the ULISSE data due to the low angular resolution although it is more sensitive than the second one for the Tenerife data. In relation to the ULISSE data the method based on the number of pips above different thresholds gives a similar result as the one obtained applying the standard method of data analysis [5]. This could be possibly due to the large sky area covered (bigger than the area covered by the Tenerife experiement as shown in figure 1) which favours the ergodicity of the CMB temperature field.

References

1. Smoot, G. F., et al. ApJ, 396, L1 (1992).
2. Meyer, S.S., Cheng, E.S. and Page, L.A. ApJ, 371, L7 (1991).
3. Ganga, K., Cheng, E., Meyer, S. and Page, L. ApJ, 410, L57 (1993).
4. Watson, R.A., Gutiérrez de la Cruz, C.M., Davies, R.D., Lasenby, A.N., Rebolo, R., Beckman, J.E. and Hancock, S. Nature, 357, 660 (1992).
5. De Bernardis, P., Masi, S., Melchiorri, F., Melchiorri, B. and Vittorio, N. ApJ, 390, L57 (1992).
6. Abbott, L.F. and Wise, M.B. ApJ, 282, L47 (1984).
7. Cayón, L., Martínez-González, E. and Sanz, J.L. Mon. Not. astr. Soc., 253, 599 (1991).
8. Gutiérrez de la Cruz, C.M., Martínez-González, E., Cayón, L., Rebolo, R. and Sanz, J.L. Mon. Not. astr. Soc., to be published (1993).
9. Zabotin, N.A. and Nasel'skii, P.D. Sov. Astr., 29(6), 614 (1985).
10. Sazhin, M. V. Mon. Not. astr. Soc., 216, 25p (1985).
11. Bond, J.R. and Efstathiou, G. Mon. Not. R. astr. Soc., 226, 655 (1987).
12. Vittorio, N. and Juszkiewicz, R. ApJ, 314, L29 (1987).
13. Coles, P. and Barrow, J.D. Mon. Not. astr. Soc., 228, 407 (1987).
14. Martínez-González, E. and Sanz, J.L. Mon. Not. astr. Soc., 237, 939 (1989).
15. Coles, P. Mon. Not. R. astr. Soc., 234, 509 (1988).
16. Gott, J. R., et. al. ApJ, 352, 1 (1990).

Telling Adiabatic Perturbations from Gravitational Waves and the CMB Polarization

M. V. Sazhin, N. Benítez

[1] Sternberg Astronomical Institute, 119899 Moscow, Russia
   E-mail: sazhin@sai.msk.su
[2] Departamento de Física Moderna, Univ. de Cantabria, Santander, Spain

1 Introduction

During the 1992 two groups of observers announced the detection of the CMB large scale anisotropy [1 -5]. They claim to have observed at least the quadrupole anisotropy, with a value of $\sim 10^{-5}$ in the case of Relikt and $\sim few \cdot 10^{-6}$ for COBE. The CMB anisotropy is one of the most effective tools for the investigation of the Universe. The large scale fluctuations which cause the large scale anisotropy are defined by super high energy physics- namely the Grand Unification Theory (GUT)- as was clarified by the theory of inflation [6-9]. Therefore, the value of the anisotropy lower spherical harmonics allows us to make some conclusions about the physics of the early Universe and thus, about the physics of particle interactions at high energies.

Two types of metric fluctuations are generated at the inflation epoch: adiabatic perturbations (below AP) and tensor perturbations (or gravitational waves, below GW). The amplitudes of AP and GW are defined by the different parameters of GUT. So, it is very important to discriminate between the AP and GW contribution to the CMB anisotropy in order to obtain more information about the physics of the early Universe. As we consider, this can be achieved through the observation of both CMB anisotropy and polarization.

The CMB anisotropies become lineary polarized when they are scattered by the primordial plasma free electrons. The degree of polarization is not too big: the maximal value of the polarized component reaches 20%.

Three conditions are necessary in order to generate polarization:

- CMB anisotropy at the moment of scattering
- Dense plasma which can provide a big number of scattered events
- The expansion of the Universe. This creates a rapid decrease in the density of free electrons that makes negligible the probability of a second scattering.

Therefore, the polarization of the CMB can be created during two epochs:

- at the main recombination which took place in at $z \approx 1000$

and

– at the second recombination epoch which could take place later ,for instance, at $z \approx 100$, or when the first objects in the Universe were created

AP and GW add in different ways to the CMB anisotropy: for standard models, the GW contribution to the quadrupole is as small as $5 - 20\%$ of that of AP [11,12]. This is due to the fact of AP and GW amplitudes being determined by different parameters of the early Universe. For example, in the simplest case of scalar potential driving inflation

$$V(\phi) = \frac{\lambda}{4}(\phi^2 - \phi_0^2)^2$$

the AP low wavelength amplitudes are determined by $\sqrt{lambda}$ and the GW amplitudes are determined by $\frac{\lambda}{4}\phi_0^4$. Therefore, it is important to measure the AP and GW amplitudes separately in order to reconstruct the shape and parameters of the field which drives inflation.

Here it will be shown that the amplitudes of anisotropy

$$anisotropy = k_{11}A^2 + k_{12}B^2$$

and polarization

$$polarization = k_{21}A^2 + k_{22}B^2$$

are determined by a set of linear equations with coefficients k_{ij} whose determinant is not equal to zero. Therefore, we can resolve this system, find separately the AP and GW amplitudes A^2 and B^2 and thus, determine the parameters of the field driving inflation.

2 Basic equations

We assume a Friedman model of the Universe with $\Omega = 1$, $\Lambda = 0$. and the dust-like state equation $p = 0$.

Taking into account metric fluctuations we have the following expression for the linear interval

$$ds^2 = a^2(\eta)\left[d\eta^2 - (\delta_{\alpha\beta} + h_{\alpha\beta})dx^\alpha dx^\beta\right]$$

where the small values $h_{\alpha\beta}$ satisfy the Einstein equations. E.M. Lifshitz [16] showed that they can be divided into three independent types: scalar, vectorial, and tensor perturbations. In this work, we shall study the effects on the CMB of only the first and third type, and more specifically, the growing modes of adiabatic perturbations (AP) and cosmological gravitational waves (GW).

After performing the fourier transformation of the metric perturbations in the usual way we choose the following gauge for SP:

$$h_{\alpha\beta}(k,\eta) = h_A(k)(k\eta)^2\gamma_\alpha\gamma_\beta$$

and for GW

$$h_{\alpha\beta}(k,\eta) = (h_+t_{\alpha\beta} + h_\times s_{\alpha\beta})\frac{J_{3/2}(k\eta)}{(k\eta)^{3/2}}$$

where $h_A = C_A(k)A_A(k)$, $h_+ = C_+(k)A_{GW}(k)$ and $h_\times = C_\times(k)A_{GW}(k)$. The $C(k)s$ are random spectral factors obeing delta-function correlation properties for k and the $A(k)s$ are functions of k. Here we wrote only those terms needed in future calculations. The symbols $+$ and $\times$ correspond to each one of the GW polarization states. The expressions for $t_{\alpha\beta}$ and $s_{\alpha\beta}$ can be found in .

In future calculations it is assumed a Harrison - Zeldovich spectrum for the GW

$$A_{GW} = Ak^{-3/2}$$

In the case of AP we choose a nearly flat spectrum

$$A_{AP} = B(k)k^{-3/2}$$

where B shows a logarithmical dependence on k [10].

For stochastic perturbations, the corrections to the metric will be, in general, a tensor random field and the spectral densities will be stochastic functions therefore. We consider that both GW and SP are distributed homogeneously and isotropically, which means that the spectral densities will depend only on the module of the wave vector. Besides, each kind of perturbation, the ones corresponding to the different states of GW polarization $C_+(k)$, $C_\times(k)$ as well as to AP $C_A(k)$, is not correlated to the others.

In order to describe the CMB, we follow [14] and introduce the symbolic vector $\hat{n}$. Here $\hat{n}$ is a kinetical distribution function which depends on the coordinates x_α and the momenti p_α of the photons. Its components are n_l , n_r -the occupation numbers of photons with self perpendicular states of polarization- and n_u, which describes the correlation between n_l and n_r. The full photon occupation number is $n = n_l + n_r$. The CMB anisotropy is given by the fluctuations of this value. We denote them with the help of the vector $\hat{\delta}$. The kinetical equation for $\hat{\delta}$ in the Rayleigh-Jeans part of the spectrum is of the form

$$\frac{\partial\hat{\delta}}{\partial\eta} + \frac{\partial\hat{\delta}}{\partial x^\alpha}e^\alpha = \frac{1}{2}\frac{\partial h_{\alpha\beta}}{\partial\eta}e^\alpha e^\beta - \sigma_T N_e a(\eta)\left\{\delta - \frac{1}{4}\int\int \hat{P}(\Omega,\Omega')\hat{\delta}(\Omega')\,d\Omega'\right\} \quad (1)$$

where $e^\alpha = (\cos\phi\sin\theta, \sin\phi\sin\theta, \cos\theta)$

The expression of the scattering matrix P through the radiation angles of incidence θ and ϕ and detailes about equation one can be found in [13, 14, 15].

3 The plane wave solution

As it will be shown later, in order to find the anisotropy and polarization of the CMB for a random spectrum of perturbations, it is enough to solve the transfer equation(1) for the plane wave case. This also helps to make clear the physical meaning of our calculations.

We choose the coordinate system so that the perturbation propagates along the $z-$ axis. The angle between the phonton propagation axis and the $z-$axis is θ. Then we have the following expressions for the first term on the right of equation (1)

$$F_{GW} = (h_+ \cos 2\phi + h_x \sin 2\phi) \; fracJ_{5/2}(k\eta)(k\eta)^{3/2}(1 - \mu^2)$$

$$F_{SP} = h_A k^2 \eta (\mu^2 - 1/3)$$

h_+, h_x and h_A are the constant amplitudes of GW and AP respectively. We deleted a monopole component of the anisotropy in order to have the second Legendre polinomy. Out of convenience, we choose the form for $\hat{\delta}$ as in [14,15].

$$\delta(k, \eta, \mu) = \alpha_A(k, \eta, \mu)(\mu^2 - \frac{1}{3}) \begin{pmatrix} 1 \\ 1 \\ 0 \end{pmatrix} + (\alpha_+ \cos 2\phi + \alpha_x \sin 2\phi)(1 - \mu^2) \begin{pmatrix} 1 \\ 1 \\ 0 \end{pmatrix} +$$

$$+\beta_A(1-\mu^2) \begin{pmatrix} 1 \\ -1 \\ 0 \end{pmatrix} + \beta_+ \begin{pmatrix} (1+\mu^2) \cos 2\phi \\ -(1+\mu^2) \cos 2\phi \\ 4\mu \sin 2\phi \end{pmatrix} + \beta_x \begin{pmatrix} (1+\mu^2) \sin 2\phi \\ -(1+\mu^2) \sin 2\phi \\ -4\mu \cos 2\phi \end{pmatrix} \quad (14)$$

After sustituting this vector in the transfer equation (1) we obtain a set of integro-differential equations.

The polarization of the CMB is accumulated mainly during the recombination. The optical depth during this period is

$$\tau(\eta) = \int_\eta^\infty \sigma_T N_e a(\eta') \, d\eta'$$

This formula can be written as

$$\tau = A\eta^{-3} \exp(-2\eta^2/\Delta\eta_r)$$

where $A = 7 \cdot 10^5 \cdot \Delta\eta_r = 0.1$ and the recombination time $\eta_r = 1$ corresponds to $z \approx 900$. $\Delta\eta_r$ is the interval during which the optical depth due to Thomson scattering changes from 1 to 0.

We shall make our calculations for $\eta > \eta_r$ in the approximation $\alpha \gg \mu k\alpha$ and $\beta \gg \mu k\beta$. It is easy to see that this is equivalent to $k\Delta\eta_r \ll 1$. This means that the functions α, β depend only on k and η, and we can write our integro-differential system as

$$\frac{d\alpha}{d\eta} = F(k, \eta) - \frac{9}{10}\sigma_T N_e a(\eta)\alpha - \frac{6}{10}\sigma_T N_e a(\eta)\beta$$

$$\frac{d\beta}{d\eta} = -\frac{1}{10}\sigma_T N_e a(\eta)\alpha - \frac{4}{10}\sigma_T N_e a(\eta)\beta$$

We wrote only one pair of equations as they are identical for both GW and AP. Introducing the function $\xi = \alpha + \beta$ and adding these equations we obtain a simple system of differential equations. The analitical solution of this system has the form

$$\xi = \int_0^\eta F(k, \eta') exp(-\tau(\eta, \eta')) \, d\eta'$$

$$\beta = -\frac{1}{10} \int_0^\eta \sigma_T N_e a(\eta) F(\eta') exp(-\frac{3}{10} \tau(\eta, \eta')) \, d\eta'$$

where τ is optical depth from η to η' For both AP and GW we shall present $F(k, \eta)$ as an expansion in a Taylor series. Solving numerically the above equations one can find α and β from them. The solutions show that the contribution of AP and GW to the anisotropy and the degree of polarization are different.

We got the degree of polarization and anisotropy of the CMB at the end of recombination. In order to find them at the present moment we can no use anymore our integro-differencial equations. The anisotropy can be find in the standard way. In the case of polarization, and assuming that there is no reonization, that is, taking $N_e = 0$ after the end of recombination:

$$\beta_{present} = \beta_{recombination} \exp(-ik(\eta_{present} - \eta_{recombination}))$$

The specific goal of our talk is the estimation of the level of the CMB polarization. Using the above equation we can estimate the r.m.s degree of polarization as being

$$p_{r.m.s.}^2 = 1.5 \cdot 10^{-4} \cdot A_{SP}^2 + 2 \cdot 10^{-6} \cdot A_{GW}^2$$

The anisotropy induced by GW and SP has been found by many athours. We will use equations from [11]

$$\sigma^2 = 5.2 \cdot 10^{-3} A_{SP}^2 + 4.5 \cdot 10^{-5} A_{GW}^2$$

The determinant of this system is $8.6 \cdot 10^{-8}$. As we wanted to show, this means that the amplitudes of AP and GW can be estimated separately.

We have estimated the total anisotropy and polarization in a wide angular range. The same situation is valid for some other specific angular range. Therefore, our general conclusion about the possibility of separation of SP and GW is valid.

References

[1] Strukov,I.A., Brukhanov,A.A., Skulachev,D.P., Sazhin,M.V., Pis'ma v Astron. Zh. **18** (1992) 387 (Sov.Astron.Lett.,**18**, 153,(1992))

[2] Strukov,I.A., Brukhanov,A.A., Skulachev,D.P., Sazhin,M.V., Mon. Not. R. astr. Soc. **258** (1992) 37P

[3] Smoot, G. et al., Astrophys. J. **396** (1992) L1.

[4] C.L.Bennett,C.L. et al., Astrophys. J., **396** (1992) L7

[5] Wright et all., Astrophys. J. **396** (1992) L13

[6] Inflationary cosmology (1986),Ed.Abbott,L.F. and So-Young Pi, World Sci.

[7] Linde, A., Particle Physics and Inflationary (1990) Cosmology, Harwood.

[8] Dolgov, A.D., Sazhin, M.V. & Zeldovich, Ya.B., Basic of Modern Cosmology (1990) Editions Frontiers.

[8] Rubakov, V., et all., Phys. Lett. **115B** (1982) 189

[9] Starobinsky A., Phys. Lett., **117B** (1983) 175

[10] Starobinsky, A.A., Pisma v AZ **9** (1983) 579

[11] Starobinsky, A.A., Pisma v AZ **11** (1985) 394

[12] Kaiser, N. , Mon. Not. R. Astron. Soc., **202** (1983)

[13] Chandrasekhar, S., Radiative Transfer. (1960) Dover Publ.

[14] Basko M.M., Polnarev A.G., Mon. Not. R. Astron. Soc. **191** (1980) L47

[15] Polnarev A.G., Astron. Zh. **62** (1985) 1041

[16] Lifshitz E.M., JETP, **16** (1946)

Imprints of Galaxy Clustering Evolution on the CMB

Enrique Martínez-González and Jose L. Sanz

Dpto. de Física Moderna, Universidad de Cantabria and Instituto de Estudios Avanzados en Física Moderna y Biología Molecular, CSIC – Universidad de Cantabria, Facultad de Ciencias, Avda. Los Castros s/n, 39005 Santander, Spain.

1 Introduction

Cosmic Microwave Background (CMB) photons are redshifted by the gravitational potential of matter inhomogeneities present in the universe, in addition to the cosmological redshift due to the expansion of the universe. Gravitational redshift affects the photons all the way from recombination till now. Most of the effect comes from density perturbations of small amplitude which evolves in the linear regime. They produce fluctuations in the temperature of the microwave photons that where first calculated by Sachs and Wolfe (1967) [1] in the case of an Einstein–de Sitter universe. These fluctuations dominate the large angular scale anisotropy of the CMB temperature.

However, at the late stages of the evolution of the universe when clusters and superclusters of galaxies start to form, the matter enters in a phase of non-linear evolution producing a gravitational potential varying with time. The formalism to calculate the temperature anisotropies generated in this non-linear phase has been developed in [2]. Analytical calculations of the anisotropies produced by massive structures and voids [3] indicate levels of anisotropies of $10^{-6} - 10^{-5}$ and $10^{-7} - 10^{-6}$ respectively. Calculations using N–body simulations confirm that the effect is relevant for large positive density perturbations but not for the negative ones [4].

In perturbation theory within a flat universe the first order density perturbations gives a null contribution to the temperature fluctuations due to the time varying potential. The first non-null contribution comes from second order perturbations. Second order density perturbations contribute an effect of the order of 10^{-6} for popular models of galaxy formation [5].

In the next section we will show the formalism and in section 3 it will be applied to the case of the evolution of non-linear galaxy clustering. We will assume that matter follows the galaxy distribution except for a constant bias of 2. The result is a temperature anisotropy of amplitude $\sim 10^{-5}$ on angular scales of a few arcmin. These theoretical values should be compared with the most stringent observational limits on $\Delta T/T$ at small angular scales, given by the following experiments: ATCA [6] and Fomalont et al. (1993) [7] with upper limits of 0.9 and 1.9×10^{-5}, respectively, at an angular scale of $\approx 1 arcmin$, and OVRO experiment [8] working at $7.15 arcmin$ and providing a limit of 1.5×10^{-5}.

2 Non-linear Density Perturbations

In the non-linear regime gravitational potentials evolve with time, contrary to the linear regime where they are static for a flat background. Considering scales much smaller than the horizon and non-relativistic speeds it is possible to derive the following result for the temperature fluctuations produced by non-static potentials in a flat universe [2] (other terms are negligible)

$$\left(\frac{\Delta T}{T}\right)_{NL} = 2 \int_r^o dx \cdot \nabla \varphi(t, x) \quad , \nabla^2 \varphi(t, x) = 6a^{-1} \Delta(t, x) \quad , \tag{1}$$

$\varphi(t, x)$ is the gravitational potential of non-linear density perturbations, $x(t, n) = \lambda(a)n$, $\lambda(a) \equiv 1 - a^{1/2}$ with a being the scale factor of the universe normalized to the present time ($a_0 = 1$) ($c = 8\pi G = 1$ and the horizon distance at present is $d_{H_o} = 3t_0 = 1$). $\lambda(t)$ is the distance of the photon to the observer at cosmic time t in comoving coordinates normalized to the present.

The temperature correlation function $C(\alpha)$ is obtained form equation (1) by averaging over all direction pairs separated an angle α

$$C(\alpha) = 4 \int_0^1 d\lambda_1 (n_1 \cdot \nabla_1) \int_0^1 d\lambda_2 (n_2 \cdot \nabla_2) < \varphi(\lambda_1, x_1) \varphi(\lambda_2, x_2) > \quad , \tag{2}$$

$n_1 \cdot n_2 = \cos \alpha$, ∇_a is the gradient in the direction n_a ($a = 1, 2$). Thus, $C(\alpha)$ is given in terms of the correlation of the gravitational potential at two positions x_1 and x_2 and at two different cosmic times λ_1 and λ_2. Although we do not have information on the potential correlations at different times we can relate them to the non-linear matter density corretions which are limitted to short scales ($\leq 5h-1Mpc$) and then these last correlations are restricted to spatial points at the same cosmic time. In the next section we will make use of this feature to calculate $C(\alpha)$ for several models for the evolution of galaxy clustering.

3 Models for the Evolution of Galaxy Clustering

Observations of the galaxy distribution are limitted to the local universe with approximately the same cosmic time for all galaxies. In order to parametrize the evolution of the two-point galaxy correlation function $\xi(z, r)$ with redshift we will assume it as a separate fuction of reshift and distance with the spatial dependence which is observed today: $\xi(r, z) = (r/r_0)^{-\gamma}(1 + z)^{-(3+\epsilon)}$, $\gamma \simeq 1.8$. The parameter ϵ accounts for the different type of evolution: $\epsilon = 0$ corresponds to stable clustering where clusters are already formed and relaxed at a high redshift and $\epsilon > 0$ will refer to models where clusters form more recently and are still evolving at present. The greater the value of ϵ the faster is the evolution and the younger is the age of the clusters. N-body simulations [9] predict $0 \leq \epsilon \leq 3$

for scenarios with spectral indexes in the range $1 \geq n \geq -3$. The generally most reasonable value of $n = -1$ for the nonlinear regime would predict $\epsilon \approx 1$. In terms of λ and comoving distance x we have

$$\xi(\mathbf{x}, \lambda) = (\frac{x}{x_0})^{-\gamma}(1 - \lambda)^{2(3+\epsilon-\gamma)}. \tag{3}$$

Relating the potential correlations to matter density correlations by considering the Poisson equation in equation (2) we can arrive to the following expression for the temperature anisotropies

$$\begin{aligned}
C(\alpha) = 48 \cos \alpha \int_0^1 \frac{d\lambda_1}{(1-\lambda_1)^2} \int_0^1 \frac{d\lambda_2}{(1-\lambda_2)^2} \left[-\int_0^x dy y \xi(\lambda_1, y) + \right. \\
\left. \frac{1}{x^3} \int_0^x dy y^4 \xi(\lambda_1, y) \right] + 72 \sin^2 \alpha \int_0^1 \frac{d\lambda_1 \lambda_1}{(1-\lambda_1)^2} \int_0^1 \frac{d\lambda_2 \lambda_2}{(1-\lambda_2)^2 x^5} \\
\int_0^x dy \xi(\lambda_1, y)(x^2 y^2 - y^4) \ ,
\end{aligned} \tag{4}$$

where $x^2 = \lambda_1^2 + \lambda_2^2 - 2\lambda_1\lambda_2 \cos \alpha$. In the previous equation it has been assumed that the matter two-point correlation function is only a function of a single time λ_1.

By using equation (4) we have estimated temperature anisotropies for several models of clustering evolution. We find that $C(0) - C(\alpha)$ takes the maximum value at $\approx 5, 2$ and 1×10^{-10} for $\epsilon = 0, 1$ and 2 respectively. This maximum difference in the correlation is attained at an angular scale of a few arcmin, with the scale of the maximum growing with ϵ. This behaviour is a consequence of the later collapse of the clusters for bigger epsilon and therefore of the bigger angle subtended by them in models with faster evolution.

4 Conclusions

We have shown that galaxy clustering evolution can have an observable effect on the CMB anisotropy on arcmin scales. The anisotropy is generated at redshifts between $1 - 3$ and therefore is not erased by reionization processes that could take place in the universe at high redshift.

The fluctuation in temperature is negative towards the center of the cluster, like for the Sunyaev – Zeldovich effect in the Rayleigh – Jeans part of the spectrum. The sign of the fluctuation is of course opposite to the S – Z effect in the Wien region.

Finally, observational limits on the anisotropies on arcmin scales [7, 6, 8] constrain the models for the evolution of galaxy clustering and therefore the cluster formation. Thus, we obtain that the stable clustering model, $\epsilon = 0$, produce anisotropies which are slightly above present upper limits whereas models with faster evolution, $\epsilon \geq 1$, do not really contradict them.

References

1. Sachs, R.K. & Wolfe, A.N. ApJ, 147, 73 (1967).
2. Martínez-González, E., Sanz, J.L. & Silk, J. ApJ, 355, L5 (1990).
3. Martínez-González, E. & Sanz, J.L. Mon. Not. astr. Soc., 247, 473 (1990).
4. Van Kampen, E. & Martínez-González, E. in Proceedings of Second Rencontres de Blois "Physical Cosmology", 582 (1991).
5. Martínez-González, E., Sanz, J.L. & Silk, J. Phys. Rev. D, 46, 4193 (1992).
6. Subrahmanyan, R., Ekers, R.D. Sinclair, M. & Silk, J. Mon. Not. astr. Soc., 263, 416 (1993).
7. Fomalont, E.B., Lowenthal, J., Partridge, R.B. & Windhorst, R.A. ApJ, 404, 8 (1993).
8. Readhead, A.C.S., Lawrence, C.R., Myers, S.T., Sargent, W.L.W., Hardebeck, H.E., & Moffet, A.T. ApJ, 346, 566 (1989).
9. Melott, A.L. ApJ, 393, L45 (1992).

Analysis of Texture on Cosmic Background Maps*

*Vahe G. Gurzadyan[1] and Sergio Torres[2]***

[1] Dept.of Theoretical Physics, Yerevan Physics Institute, 375036, Yerevan, Armenia
[2] International Center for Relativistic Astrophysics. Dip. di Fisica, Università di Roma 1. Pssa Aldo Moro 5, 00185 Rome, Italy

1 Introduction

Due to the early origin of the cosmic microwave background radiation (CMB), this signal might be the best probe we have into the early universe, and it is a powerful test for cosmological models. The observed anisotropies by the Differential Microwave Radiometer on board the Cosmic Background Explorer (COBE) [1] can be produced by a number of sources that go from local, to astrophysical, to cosmological effects, among which the most relevant are: emission from the Moon, the Earth, other planets and the Galaxy; interaction of CMB photons with electrons in hot clouds and with cosmic rays; gravitational fluctuations produced by voids and great attractors; the Doppler effect due to the observer's motion relative to the CMB; mixing of geodesic flow; gravitational waves; cosmic strings; gravitational potential fluctuations on the last scattering surface (the Sachs-Wolfe effect [2]); and gravitational lensing. Clearly the complexity of the problem of determining the cause of CMB anisotropies calls for new analysis techniques. Several statistics have been proposed [3, 4, 5, 6, 7, 8, 9] as topological descriptors of the CMB maps: hot spot number density, contour length, contour curvature or "genus", number of upcrossings, area and eccentricity of hot spots, and Euler-Poincaré characteristic. The study of these topological descriptors is based on the well established theory of the geometric properties of excursion sets of random fields [10].

According to the theory of geometric properties of random fields, in the case of two dimensional isotropic, stationary Gaussian random fields, primordial fluctuations would exhibit a clear signature. Thus the appearance of systematic effects in the data and of non Gaussian fluctuations can in principle be tested by analysing geometric properties.

Superposed to the CMB anisotropies expected from the Sachs-Wolfe effect, there may also appear the signature of the mixing of geodesics effect [11, 12].

---

* Presented by S. Torres at the Workshop on Present and Future of the Cosmic Microwave Background, Santander, June, 1993

** On leave from Universidad de los Andes, and Centro Internacional de Física, Bogotá, Colombia. E-mail: 40174::torres, torres@celest.lbl.gov

Because of the appealing possibility of extracting information about the curvature of the universe from the latter effect, we will deal with it in some detail. A consequence of mixing phenomena in geodesics propagation is the appearance of highly distorted anisotropy spots in cosmic background maps [14]. However, confusion due to galactic contamination at high latitudes and similar elongated patterns produced by instrumental noise makes this effect very difficult to observe.

Below, we present an analysis of texture on digitised COBE maps with the purpose of obtaining the elongation properties of structures and to compare the peak statistics in the data with the numbers expected from density fluctuations in the early universe.

2 Effect of mixing of geodesics flow

The idea of the effect of instability of trajectories of freely moving particles can be most clearly demonstrated via the Jacobi equation written for spaces with constant curvature k:

$$d^2\mathbf{n}/d^2\lambda + k\mathbf{n} = 0, \qquad (1)$$

describing the behavior of the vector of deviation, $\mathbf{n}$, of close geodesics. Solutions of this equation are determined by the sign of the curvature: when $k < 0$ one has exponentially deviating geodesics.

A more rigorous treatment [11, 12] includes the study of the projection of geodesics from (3+1)-dimensional Lorentzian space to a 3D Riemannian one, and the behavior of time correlation functions for geodesic flows on homogeneous isotropic spaces with negative curvature. Geodesic flows, being Anosov systems (locally if the space is not compact), are exponentially unstable systems possessing: the strongest statistical properties (mixing), non-zero Lyapunov characteristic exponents, and positive Kolmogorov-Sinai (KS) entropy h [15].

For Anosov systems two geodesics in 3-space deviate exponentially according to the law

$$\mathbf{n}(\lambda) = \mathbf{n}(0)\exp(\chi\lambda), \qquad (2)$$

where χ is the Lyapunov exponent.

For a homogeneous isotropic Friedmannian Universe with $k = -1$ the Lyapunov exponent is determined by the only parameter a, the diameter of the Universe:

$$\chi = 1/a, \qquad (3)$$

while Lyapunov exponents vanish when $k = 0, +1$.

Time correlation functions, describing the decrease of perturbations also decay exponentially as determined by KS-entropy: $h = 2\chi$.

For the Universe expanding as

$$a(t) = a(t_0)(t/t_0)^\alpha, \qquad (4)$$

the relation between the quantitative measurement of the distortion of patterns, ϵ, and Ω is given by [14]

$$\frac{\ln(1/\epsilon)}{(\frac{1-\Omega_0}{1+z\Omega_0})^{\frac{1}{2}}} = \begin{cases} \alpha/(1-\alpha)[1-(1+z_1)^{1-1/\alpha}] & \alpha < 1 \\ \ln(1+z_1) & \alpha = 1 \end{cases} \tag{5}$$

where Ω_0 is its present value ($z = 0$), z_1 corresponds to the time when matter becomes non relativistic and $z(\approx 1000)$ to the decoupling time.

The parameter of elongation ϵ defined via the divergence of geodesics in (3+1)-space:

$$\epsilon = \frac{l(t)}{l(0)}, \quad l(t) = l(t_0)\frac{a(t)}{a(t_0)}\exp(\chi\lambda(t)), \tag{6}$$

appraoches 1 when Ω tends to 1 as shown in Fig. 1.

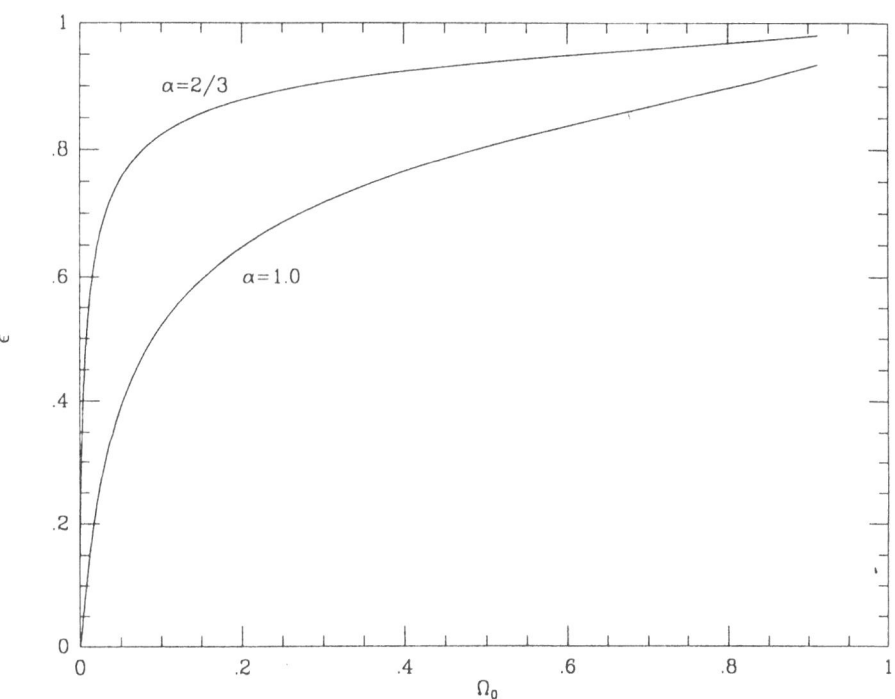

Fig. 1. Eccentricity parameter ϵ as a function of Ω for two exponents for the law of the expansion rate of the Universe, $\alpha = 1$ and $\alpha = 2/3$.

The typical pattern of a hot spot as seen today in a $k < 1$ universe would exhibit a very complex shape. The elongation, ϵ, measures the smallest-to-largest ratio of diameters of one-connected regions and is related to the "degree of complexity" [14] of anisotropies.

3 Preparation of Data

COBE's Differential Microwave Radiometers (DMR) have mapped the CMB at three frequencies: 31, 53, and 90 GHz [17]. The analysis presented here is based on DMR's results of the first year of observation [1]. The three sum (A+B) DMR maps were digitized on pixels of $\approx 2.9°$ to agree with the quadrilateralized spherical cube data base used by DMR [16]. These maps have the dipole anisotropy removed and have been smoothed. The areas of the sky for $140° < \ell < 220°$ and $|b| > 74°$ were excluded because of the high distortion of shapes near the edges of the maps due to the projection on a plane. b and ℓ are galactic latitude and longitude respectively. The systematic contribution of the Galaxy has been cut by excluding the band $|b| < 20°$ as done in the analysis of DMR's data [18]. The digitized data is then projected back to a 6144 pixel cube (1024 pixels per face) in which the features of the original maps are recovered free of distortion. The errors in the temperature and angle introduced by the digitization process are estimated to be $\approx 54\,\mu K$ and $\approx 1.4°$. The limiting angular resolution is dictated by the pixel size which in any case is comparable with DMR's angular resolution of $3.2°$.

Due to the limitations in the digitization process, only anisotropy spots at one threshold level were available for our analysis. The threshold level ν, is defined as the sky temperature in units of the rms fluctuation level, $T_\nu = \nu\sigma$. The ν values of the hot/cold spot analysis presented here are ± 1, ± 1.4, and ± 1 for the 31, 53, and 90 maps respectively. Negative thresholds define cold spots, that is, areas formed by pixels with temperatures less than $-\nu$.

4 Analysis of DMR Maps

Program PATREC, a pattern recognition algorithm [19] was used on the maps prepared by the above mentioned procedure to identify hot and cold spots and analyze their shape and area. A hot spot is defined as a continuous region of the map formed by the pixels whose temperature is higher than a preset threshold ν. To identify hot spots on the maps, the algorithm compares an elliptical test pattern with all possible areas on the sky-map of the same size. The distribution of the eccentricity parameter $\epsilon = r_{min}/r_{max}$ of those patterns actually identified is used to characterize the shapes of the structures. Limiting the test pattern to have an elliptical shape tends to underestimate the sizes of hot spots, and results in inefficiencies of the algorithm in finding spots. The efficiency depends on the threshold level. By a more direct method to find hot spots it was found that at the threshold levels in our analysis the error in spot number counts is no larger than 3%. This more direct method consists of forming tree data-structures with all lit pixels as nodes. The number of spots equals the number of trees.

Even though it is well known that the 1 year DMR maps have a noise level that is comparable with the measured anisotropies $\Delta T/T \approx 6 \times 10^{-6}$ it is important to notice that some of the hot spots on the maps can indeed be in the CMB, assuming the data is not dominated by systematic errors, which is a reasonable assumption due to the careful treatment of the subject by the COBE-DMR group [22]. Detection of anisotropies by other instruments [13], and new COBE maps with less noise will allow to firmly establish the nature of CMB anisotropies.

Table 1 shows the hot and cold spot counts along with their mean area and eccentricity. The mean value of ϵ for the combined maps is 0.6 indicating the presence of elongated patterns.

Table 1. Number of hot plus cold spots N, normalised to 4π steradians, in digitised COBE maps and mean values $\pm 1\sigma$ of number of hot plus cold spots found in simulated noise maps $\langle N_{noise}\rangle$ and 'Harrison-Zeldovich maps $\langle N_{HZ}\rangle$. Mean eccentricity $\langle \epsilon \rangle$ and area $\langle A \rangle$ are for COBE maps. Area is in squared degrees and map frequency in GHz. All topological descriptors are given for the indicated threshold level ν

Freq.	ν	$\langle \epsilon \rangle$	$\langle A \rangle$	N	$\langle N_{noise}\rangle$	$\langle N_{HZ}\rangle$
31	1.0	0.60	100	188	185 ± 14	183 ± 14
53	1.4	0.56	119	111	153 ± 9	114 ± 14
90	1.0	0.59	77	183	201 ± 15	171 ± 15

In order to study the effects due to instrumental noise and COBE's sky coverage, noise maps were generated [20] by simulating COBE's polar orbit and assigning to each pixel in the sky a temperature equal to a random number extracted from a Gaussian distribution. The variance of this noise term is given by the square of the corresponding DMR noise level [17] weighted by $1/N$, where N is the number of observations. The resulting map agrees with the expected fact that, due to the polar orbit, the equatorial (celestial) band should exhibit noisier behavior compared with the celestial polar caps. A $2.9°$ Gaussian smoothing was applied to each generated noise map. A total of 100 maps were used. The average number of hot plus cold spots on simulated noise maps along with their 1σ dispersion is reported in Table 1. These numbers are given for the same thresholds as those used in the DMR maps. The average eccentricity of noise maps is very close to that of DMR maps ($\langle \epsilon \rangle = 0.55$).

5 Primordial density fluctuations

Because of the Sachs-Wolfe imprint on the CMB, the fluctuation spectrum, $P(k) \sim k^n$, can be probed directly for scales larger than the horizon at time of last scattering. This can be done by means of the CMB angular correlation function. However, to test for the Gaussian character of these fluctuations it is necessary to look at the number of spots and other topological descriptors [6].

In order to compute the number density of spots expected from density fluctuations with a Harrison-Zeldovich spectrum, $n = 1$, $P(k) \propto k$, a montecarlo procedure to generate simulated maps was used. To generate simulated maps the CMB temperature was calculated from a harmonic expansion with Gaussian random harmonic coefficients $a_{\ell m}$. These harmonic coefficients have zero mean and variance $\propto 1/\ell(\ell+1)$ [21]. Their variance has been normalized to COBE's quadrupole $Q_{rms\text{-}PS} = 16\mu$ K. Superposed to the cosmic signal, the instrumental noise has been included as described in the previous section. The averages reported here are based on 100 realizations. The mean eccentricty of spots at threshold level $\nu = 1$ in these 'Harrison-Zeldovich' maps is 0.6. The mean number of cold plus hot spots in these maps and their 1σ dispersion is shown in Table 1. The number of cold plus hot spots for the montecarlo simulations quoted in Table 1 are at the same threshold level as their corresponding level in the DMR map.

6 Conclusions

It is clear that eccentricity alone is not a good discriminator for telling appart noise maps from any 'signal' map. This facts makes very difficult the identification of a clear signature for the mixing of geodesics effect. However, because of the possibility of a direct determination of Ω as a consequence of mixing, new analysis techniques should be considered. One possibility is to use the fractal dimension of hot spots as a measure of their degree of complexity. The fact that structure due to primordial fluctutions appears superposed on the anisotropies produced by the geodesic flow effect makes a strightforward interpretation of the data, in this context, more difficult. The number of spots appearing in the 53 GHz map at threshold level $\nu = 1$ is consistent with a Harrison-Zeldovich spectrum, while for the 90 GHz map it is only marginally consistent. Because of the high level of noise of the 31 GHz DMR channels, the montecarlo procedure shows that at the $\nu = 1$ threshold level the number of spots in the map does not provide usefull information.

Acknowledgements. The authors would like to thank A. A. Kocharyan and R. Ruffini for valuable discussions and the SuperComputer Computations Research Institute of Florida State University where the pattern recognition runs were made. This research is in part covered by Colciencias (the Colombian foundation for sciences) grant N. 1204-05-007-90 and the Italian Ministry of Foreign Affairs.

References

1. Smoot, G. F., et al. ApJ, 396, L1 (1992).
2. Sachs, R. K. and Wolfe, A. N. ApJ, 147, 73 (1967)
3. Sazhin, M. V. Mon. Not. astr. Soc., 216, 25p (1985).
4. Vittorio, N. and Juszkiewicz, R. ApJ, 314, L29 (1987).
5. Bond, J.R. and Efstathiou, G. Mon. Not. R. astr. Soc., 226, 655 (1987).
6. Coles, P. Mon. Not. R. astr. Soc., 234, 509 (1988).
7. Gott, J. R., et. al. ApJ, 352, 1 (1990).
8. Martinez-Gonsález, E., Cayon, L. in *The Infrared and Submillimetre Sky after COBE* (eds M. Signore and C. Dupraz) 303 (Netherlands: Kluwer Academic Press, 1992)

9. Martínez-Gonsáles, E. and Sans, J. L. Mon. Not. R. astr. Soc., 237, 939 (1989).

10. Adler, R. J. The Geometry of Random Fields, (Chichester: Wiley) (1981)

11. Gursadyan V.G., Kocharyan, A.A. in: *Quantum Gravity*, (eds Beresin V.A., Markov M.A., and Frolov V.P.) 689 (World Sci. Singapore, 1991).

12. Gursadyan, V.G., Kocharyan, A.A. Astr&Ap, 260, 14 (1992).

13. Ganga, K., Cheng, E., Meyer, S. and Page, L. ApJ, 410, L57 (1993).

14. Gursadyan, V.G., Kocharyan, A.A. preprint SISSA, Ref.216/92/A, Trieste (1992), Int.Journ.Mod.Phys. D., 2, No.1, 1993; Europhys.Lett. 22, 231 (1993)

15. Arnold, V.I. *Mathematical Methods of Classical Mechanics*, Nauka, Moscow (1989).

16. Torres, S., et al. in *Data Analysis in Astronomy III* (eds V. di Gesu, L. Scarsi, and M.C. Maccarone) 319 (New York: Plenum, 1989).

17. Smoot, G.F., et al. ApJ, 360, 685 (1990).

18. Bennett, C.L., et al. ApJ, 396, L7 (1992).

19. Torres, S. PREPRINT-ICRA-12-3-93 (1993).

20. De Greiff, A. and Torres, S. in *Escuela Nacional de Física VIII ENAFIT*, p. 303 (1992)

21. Peebles, P. J. E. ApJ, 263, L1 (1982).

22. Kogut, A., et al. ApJ, 401, 1 (1992).

Sakharov Modulation of the Spectrum of Initial Perturbations and its Manifestation in the Anisotropy of Cosmic Microwave Background and Galaxy Correlation Function

H.E. Jørgensen[1] *E.V. Kotok*[1,2,5*], *P.D. Naselsky*[1,2,4*], *I.D. Novikov*[1*,2,3,6]

[1] University Observatory, Østervoldgade 3, DK-1350 Copenhagen K, Denmark
[2] NORDITA, Blegdamsvej 17, DK-2100 Copenhagen Ø, Denmark
[3] Astro Space Center of P.N.Lebedev Physical Insitute, Profsoyuznaya 84/32, Moscow, 117810, Russia
[4] Rostov State University, Zorge 5, Rostov-Don, 344104 Russia
[5] Institute of Applied Mathematics, Miusskaya pl. 4, 125000, Moscow, Russia
[6] Theoretical Astrophysics Center, Blegdamsvej 17, DK-2100, Copenhagen Ø, Denmark

* permanent address

Abstract. We discuss the peculiar periodic dependence in the initial spectrum of the baryonic matter distribution on wavelength (Sakharov oscillations). This effect is specific for acoustic modes of perturbations in cosmic plasma in any cosmological model as a result of the change of the equation of state at the epoch of recombination. Sakharov oscillation should manifestate themselves by specific anomalies in the angular correlation function of the microwave background anisotropy on scales $10' - 2°$, in practically any cosmological model, and by analogous anomalies in the space correlation function of galaxy distributions on a scale of $30 - 150$Mpc in cosmological models with baryonic dark matter. We propose special methods for filtering of this effect in the observational data.

1 Introduction

The origin of the large scale structure in the Universe has been studied by a number of authors under various assumptions, and it still remains one of the central issues in cosmology. The recent discoveries of the Great Attractor (GA) [1], the large scale correlations in the distribution of galaxies [2] and the measurements of the anisotropy of the cosmic microwave background, (CMBR) [3]-[5], [23], [24], [41], present cosmology with a whole series of problems connected with the spectrum of density fluctuations on scales $r \simeq 10 - 100h^{-1}$Mpc[6] and larger scales [7].

The data from COBE allow one to limit the range of the possible power indices n of the power spectrum of the matter density perturbations $P(k) \propto k^n$ where k is a wave number. At the '1σ' level Smoot *et al.*[3] obtained $n = 1.15^{+0.45}_{-0.65}$ which is very close to the Harrison-Zel'dovich slope of $n = 1$ on the scales $k \leq 3 \cdot 10^{-3}h$Mpc^{-1}. However, is this the only possible interpretation of the COBE data? In principle it is possible that the spectrum of fluctuation on other scales was more complex than just a simple power law.

Recent large scale surveys as the APM [2], QDOT [7], [8], the CfA [9] and the IRAS 1.2 Jy survey [10] allow one to obtain the power spectrum of density fluctuations at large scales. The results are well fitted by the analytic approximation suggested by Peacock [11]:

$$P(k) \propto A_0 \Omega^2 k \left[1 + (k/k_c)^q\right]^{-1}; \qquad 0.03 < kMpc/h \le 1.0$$

where $A_0 \simeq 1.6 \cdot 10^{-10}$; $q = 2.4$ and $k_c = 0.024h\text{Mpc}^{-1}$; Ω is the total density in the Universe relative to the critical density; $h = H_0/(100\text{km}\cdot \text{s}^{-1}\cdot\text{Mpc}^{-1})$. Note, that on smaller scales the processes of non-linear gravitational clustering of matter transformed the primordial form of the power spectrum and therefore complicate the interpretation of observations. What kind of spectrum is at greater scales (smaller than k_c)? We'll focus attention on the range

$$3 \times 10^{-3}h\text{Mpc}^{-1} \le k \le (3 \times 10^{-2} - 10^{-1})h\text{Mpc}^{-1}.$$

To analyze the problem we divide it into two parts. In the first part we will discuss the reasons of the non-monotonical character of the spectrum, and in the second one we will propose methods for the experimental investigations of k in the range mentioned above. We begin with some general reasoning (see [25], [28], [38]).

It is well known that at the time of recombination ($z_{rec} \simeq 1100$) there existed acoustic modes of perturbations in the cosmic plasma due to interaction between matter and radiation. These modes correspond to the comoving linear scales

$$r_d \le r \le r_{acoust} = r_{rec}\frac{v_{acoust}}{c} \approx 100 - 200h^{-1}\text{Mpc}. \tag{1}$$

Here r_d is the damping scale, $r_d \sim 10h^{-1}\text{Mpc}$, and r_{rec} is the event horizon at $z_{rec} \simeq 1100$, all of them are scaled to the present epoch, $v_{acoust} = 3^{-1/2}/\sqrt{1 + 3\rho_b/4\rho_r}$ and ρ_b/ρ_r is the ratio of baryon and radiation densities at z_{rec}. The presence of acoustic modes is a natural part of the evolution of density perturbations in the hot plasma of the early Universe (Silk [13]; Peebles and Yu [14]; Zel'dovich and Novikov, [15]). They appear in numerical computations (e.g. Fukugita et al. [16]; Muciaccia et al. [17]) as well as in analytical work (Doroshkevich [18]). After recombination the motion of matter is independent of the background radiation.

Sakharov [12] was the first to point out that the transitional phenomena in the growing perturbations lead to a peculiar periodic dependence of the perturbation amplitudes on wavelength. He considered the cold model of the Universe, though, qualitatively speaking, the same phenomena occur in the hot model as a result of recombination. These Sakharov oscillations have been discussed by Peebles [42] in a framework of the baryonic model with adiabatic initial perturbations.

In nonbaryonic dark matter models (for example - CDM models) due to the peculiar gravitational field of the dark matter, baryonic density fluctuations follow those of the dark matter after the recombination, whereas the distribution of the microwave background photons gets frozen on the celestial sphere thus preserving the information about the inhomogeneities at the last scattering surface. Therefore, in the present spatial distribution of matter, which is determined mainly by that of the dark matter, acoustic modes are practically erased, while the angular distribution of CMBR has preserved the signature of the acoustic motions of matter.

Thus, if the dark matter in the Universe has a baryonic nature, then the non-monotonic character of spectrum due to Sakharov effect has to manifest itself in the spatial galaxy distribution on typical scales $r \approx r_{acoust} \equiv R$. In [38] we call these peculiarities long distance correlations. And vice versa, if the present density of non baryonic dark matter essentially exceeds ρ_b, then modulation effects in the space correlation function $\xi(r)$ is suppressed. This crucial dependence $\xi(r)$ (at $r \simeq R$) from the nature of dark matter allows one, in principle, to use Sakharov osillation as a test for determination of the parameter Ω_b and also the total matter density Ω, see [25], [28], [38]. However the observational data existing for the range $r \sim 10^2 - 2 \cdot 10^2$Mpc have considerable uncertainties. Thus, it is very important to obtain more reliable data for the distribution of galaxies on these scales, and to search for temperature fluctuations of the CMBR angular scale at intermediate $20' < \theta < 2^0$.

Unfortunately, for angles $\theta < 2^\circ$ there exist only upper limits on the anisotropy: $\Delta T/T(\theta = 4'.5) < 3 \times 10^{-5}$ (Uson and Wilkinson [19]) and $\Delta T/T(\theta = 7'.15) < 2.1 \times 10^{-5}$ (Readhead et al. [20]). For the intermediate angular scales the UCSB experiment with FWHM=$1^\circ.5$ gave $\sqrt{c_{1.5^\circ}(0)} < 1.4 \cdot 10^{-5}$ at the 95% confidence level (Gaier et al. [21]) and the MIT experiment with FWHM= $3^\circ.8$ gave $\sqrt{c_{3^\circ.8}(0)} < 1.6 \cdot 10^{-5}$ at the 92% confidence level (Meyer et al. [22]). The observational limits for angular scales $0^\circ.3 \div 1^\circ$ by Meinhold and Lubin [23], Alsop et al. [24], Gundersen et al. [41] have the level $(4 \div 5) \cdot 10^{-5}$. Because the upper limit on the scale of acoustic perturbations, r_{acoust}, is determined only by r_{rec} and $\rho_b/\rho_r(z_{rec})$, it is clear that detecting the signature of these modes in $\Delta T/T(\theta)$ would give unique information on the properties of the last scattering surface as well as the spectrum of density perturbation on scales around $r \simeq 100h^{-1}$Mpc being the most interesting from the observational point. In the paper [25] we briefly discussed this problem. Here we shall consider some aspects of these ideas more detailed.

The structure of the paper is the following. In section 2 we give the general analysis of the evolution of acoustic modes of perturbations in the CDM cosmology. Section 3 is devoted to the peculiarities of the angular correlation function of $\Delta T/T$ because of the Sakharov oscillations. In section 4 we describe the special method for filtering of the Sakharov oscillations in the angular anisotropy of the CMBR. In the section 5 we analyze briefly the Sakharov effect in the spatial correlation function of galaxies in cosmological models with different nature of dark matter. In the last section 6 we summarize our main conclusions.

2 Evolution of acoustic modes of perturbations in CDM cosmology

In this section we consider the dynamics of adiabatic perturbations of the metric, density and velocity of the plasma at the epoch of recombination. This subject was considered both analytically and numerically for the standard baryon model in earlier works [13], [14], [28], [29] as well as dark matter models [18], [27], [30], [31]. We discuss below the main stages of dynamics of perturbations in the mixture of collisionless particles (dark matter), baryons and radiation on scales where effects of photon viscosity and conductivity can be neglected.

The equations for evolution in the linear regime are [30], [32]:

$$\delta'_k + (ik\mu - \tau')\delta_k = \frac{1}{2}\mu^2 h'_k - \tau'\left[\delta_0 + \frac{1}{2}P_2(\mu)\delta_2 + \mu v_e\right],$$

$$v'_e + \left(\frac{\tilde{R}'}{\tilde{R}} - \frac{4}{3}\tau'\frac{\rho_r}{\rho_b}\right)v_e = -\frac{4}{3}\tau'\frac{\rho_r}{\rho_b}\delta_1, \tag{2}$$

$$\delta'_b + ikv_e = \frac{1}{2}h'_k,$$

$$\delta_k = \left(\frac{\Delta T}{T}\right)_k = \sum_{l=0}^{\infty}\delta_l(k,\zeta)P_l(\mu),$$

where v_e is the Fourier component of the peculiar velocity of electrons, $\mu = \cos(k \cdot l/k)$, l is the unit vector along the direction of the motion of a photon, $P_l(\mu)$ are Legendre polynomials, h_k is the metric Fourier component in the synchronous gauge, δ_b denotes perturbations in the density of matter, τ is optical depth, $\tau' = -n_e\sigma_T\tilde{R}$, $\tilde{R}$ is the scale factor, n_e - number density of free electrons, ρ_r, ρ_m are respectively densities of CMBR and baryons, $H = \tilde{R}'/\tilde{R}$ is the current Hubble parameter, 'prime' denotes derivative with respect to the conformal time, $d\zeta = dt/\tilde{R}$, and we take the speed of light $c = 1$. In order to analyze the oscillation regime of the evolution permitted by eqs.(2) we introduce the optical thickness of plasma on the scale of the wavelength λ, $\tau_\lambda = \tau'\lambda$ and we consider only modes for which $\tau_\lambda \gg 1$. Following the method of Peebles and Yu [26] we consider only the scales that are not subjected to attenuation by dissipation. For this range of k eqs.(2) become.

$$\left(1 + \frac{3}{4}\frac{\rho_b}{\rho_r}\right)\Delta''_0 + \frac{3}{4}\left(\frac{\rho_b}{\rho_r}\right)'\Delta'_0 + \frac{k^2}{3}\Delta_0 = \frac{1}{8}\frac{\rho_b}{\rho_r}\left(h''_k + \frac{\tilde{R}'}{\tilde{R}}h'_k\right),$$

$$\Delta_1 = -\frac{3i}{k}\Delta'_0 = u_e, \tag{3}$$

where $\Delta_0 = \delta_{l=0} - h''/2k^2$, $u_e = v_e + ih'_k/2k$, $h_k = A(k)k^2\zeta^2$ and $A(k)$ is the amplitude of metric perturbations. Equations (3) take a particularly simple form when $\frac{\rho_b}{\rho_r}(z = z_{rec}) \ll 1$ which is valid for $\Omega_b h^2 \ll 0.02$. In that case:

$$\Delta''_0 + \frac{k^2}{3}\Delta_0 = 0, \qquad \Delta_1 = -\frac{3i}{k}\Delta'_0. \tag{4}$$

From (4) we obtain the following expressions for Δ_0, Δ_1 which coincide with previous results (Doroshkevich [18]):

$$\Delta_0 = A(k)\cos\frac{k\zeta}{\sqrt{3}}, \qquad \Delta_1 = +\sqrt{3}iA(k)\sin\frac{k\zeta}{\sqrt{3}}. \tag{5}$$

For $\Omega h^2 \leq 0.02$ the approximate solution for acoustic modes ($k\zeta \gg 1$) is:.

$$\Delta_0(\zeta) \approx \frac{3}{8}\frac{\rho_b}{\rho_r}\frac{1}{k^2}\left(h_k'' + \frac{\tilde{R}'}{R}h_k'\right) - \frac{A(k)}{\sqrt{1+\frac{3}{4}\frac{\rho_b}{\rho_r}}}\cos\Psi(k,\zeta), \tag{6}$$

where

$$\Psi(k,\zeta) = k\int_0^\zeta \frac{d\zeta}{\sqrt{3(1+\frac{3}{4}\frac{\rho_b}{\rho_r})}}.$$

At the last scattering surface $\zeta = r_{rec}$ the phase of acoustic oscillations is given by:

$$\Psi(k,r_{rec}) = k\int_0^{r_{rec}} \frac{d\zeta}{\sqrt{3(1+\frac{3}{4}\frac{\rho_b}{\rho_r})}} = \frac{kr_{rec}}{\sqrt{3\omega}}\ln\left[\frac{2\sqrt{\omega(1+\omega+\omega\nu)}+\omega\nu+2}{2\sqrt{\omega}+\omega\nu}\right], \tag{7}$$

where $\nu = \frac{2r_{eq}}{(\sqrt{2}-1)r_{rec}}$, $\omega = \frac{3}{4(1+z)}\frac{\rho_b}{\rho_r}|_{z=z_{rec}}$, r_{eq} corresponds to the particle horizon at the redshift of equality of radiation and dark matter densities (note that $\Omega \gg \Omega_b$). For perturbations that are not affected by the dissipation effects, (2), (6) and (7) lead to the following simple expression for temperature anisotropies on the last scattering surface [29]:

$$\left(\frac{\Delta T}{T}\right)_k = -e^{-ik\mu\zeta+\tau}\int_0^\zeta d\tilde{\zeta}\tau'(\tilde{\zeta})\left[\Delta_0(\tilde{\zeta})+\mu u_e(\tilde{\zeta})\right]e^{ik\mu\tilde{\zeta}-\tau(\tilde{\zeta})}. \tag{8}$$

Since for oscillations on the considered scales, the period exceeds the characteristic timescale for change in the optical depth $\zeta \sim |\tau/\tau'|$, Δ_0 and Δ_1 can be taken outside the integral in (8) with their values at $\zeta = r_{rec}$. Furthermore, it is easy to see that the function:

$$f(\zeta) = -\tau'e^{-\tau(\zeta)}, \tag{9}$$

for the standard history of recombination [31], [35] has a Gaussian-like form and can be approximated by

$$f(\zeta) \approx e^{-\frac{(\zeta-r_{rec})^2}{2\Delta^2}}. \tag{10}$$

In the above equation (10) $\zeta = r_{rec}$ corresponds to the moment when the plasma becomes optically thin and $\Delta \simeq 0.03r_{rec}$ is the half-width of that surface [35]. Thus for acoustic modes the contribution to $(\Delta T/T)_k$ in comoving units is given by:

$$\frac{\Delta T}{T}\bigg|_{(k)} = \Delta_{rms}(z=0,\mu) \approx [\Delta_0+\mu u_e]\big|_{\zeta=r_{rec}}e^{-ik\mu(\zeta_n-r_{rec})-\frac{k^2\Delta^2\mu^2}{2}}, \tag{11}$$

where $\zeta_n \equiv \zeta \mid_{z=0}$ is the present-day horizon.

In the approximation of instant recombination ($\Delta = 0$) (7) and (11) lead to:

$$\Delta_{rms}(z = 0, \mu) = \tag{12}$$

$$= A(k) \left[3\epsilon - \frac{\cos \Psi(kr_{rec})}{(1 + \epsilon)^{1/4}} + \frac{i\sqrt{3}\mu}{(1 + \epsilon)^{3/4}} \sin \Psi(kr_{rec}) \right] \times \exp[-ik\mu(\xi_n - r_{rec})],$$

where

$$\epsilon = (1 + \nu)\frac{3\rho_b}{4\rho_r}\mid_{z_{rec}}; \ \Psi(kr_{rec}) = kR; \tag{13}$$

with R equal to the acoustic horizon at $z = z_{rec}$. Note, that to take the dissipation and non-instancy of the recombination into account it is sufficient to multiply the expression (12) by $\exp\left(-\frac{k^2\Delta^2}{2}\mu^2\right)$ (compare with (11)) and also by damping factor $\propto \exp(-\frac{k^2 l_s^2}{2})$, where l_s is the Silk's scale (see Fukugita et $al.$ [16]).

It is useful to compare our analytical results with the data from computer modelling of the processes as discussed in the paper by Muciaccia et $al.$ [17]. For this purpose we introduce the function

$$f(k) = \frac{1}{2\pi^2}k^3 \int_0^1 d\mu \langle |\Delta_{rms}(z = 0, \mu)|^2 \rangle, \tag{14}$$

which is identical to the expression for $k^3 I_c$ in the paper by Muciaccia et $al.$ [17].

On Fig.1 we plot the function $f(k)$ for the standard CDM model with the Harrison-Zel'dovich spectrum of the initial adiabatic perturbations, and with the parameters $\Omega_b = 0.03$; $h = 0.5$ and $\Omega = 1$, according to our analytical derivation. For comparison we also show the results given in Fig.1 of the paper by Muciaccia et $al.$ [17] for the analogical function, according their numerical calculations.

It is worth noting that there is a limiting case for the function (14), which corresponds to very small Ω_b. It corresponds to $\varepsilon \to 0$ in the formulae (12). For this case we obtain

$$\langle |\Delta_{rms}(z = 0, \mu)|^2 \rangle = \Phi(k) \left[\cos^2 \Psi(kr_{rec}) + 3\mu^2 \sin^2 \Psi(kr_{rec})\right] \tag{15}$$

After integration over μ the expression in brackets is equal to unity. Thus, there are no oscillations for this limiting case (see the corresponding plot on Fig.1).

3 Appearence of Sakharov oscillations in angular correlations in $\frac{\Delta T}{T}$

In this section we consider the pecularities of the angular correllations in $\Delta T/T$ at the most interesting angular scales $\theta \simeq 20' - 2°$. Following [34] we consider the correlation function

$$C(\theta, \theta_A) =$$

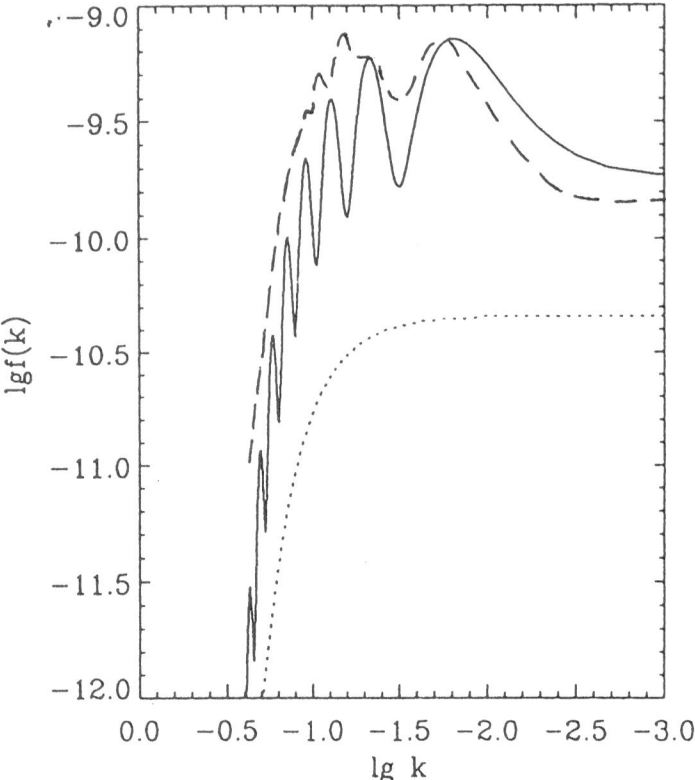

Fig. 1. $F(k)$ for the model $\Omega_R = 0.97$; $\Omega_b = 0.03$; $h = 0.5$ and the Harrison-Zel'dovich spectrum. Solid line for $\Omega_b = 0.03$, dashed-dot line for limiting case with $\epsilon \to 0$; dashed line for $k^3 I_c$ from [17].

$$= \frac{1}{4\pi^2} \int_0^\infty dk\, k^2 \int_{-1}^1 d\mu\, e^{-k^2\zeta_n^2\theta_A^2(1-\mu^2)} \langle|\Delta_{rms}(z=0,\mu)|^2\rangle J_0(k\zeta_n\theta\sqrt{1-\mu^2})$$

(16)

where θ_a is the half-width of the antenna-beam and $J_0(x)$ is the Bessel function. For the further calculation we express $J_0(k\zeta_n\theta\sqrt{1-\mu^2})$ as a series of Legendre polinomials:

$$J_0(k\zeta_n\theta\sqrt{1-\mu^2}) = \sqrt{\frac{2\pi}{k\zeta_n\theta}} \sum_{l=0}^\infty \left(2l+\frac{1}{2}\right) \frac{(2l-1)!!}{2^l l!} J_{(2l+\frac{1}{2})}(k\zeta_n\theta)P_{2l}(\mu). \quad (17)$$

Then after rather long calculations we obtain the following expression for the correlation function:

$$C(\theta,\theta_A) = \frac{1}{4\pi^2} \int_0^\infty dk\, k^2\Phi(k)\left\{\left[3\varepsilon - \frac{\cos\Psi(kr_{rec})}{4\sqrt{1+\varepsilon}}\right]^2 I_1(k,\theta,\theta_A)+\right.$$

$$+\frac{3\sin^2\Psi(kr_{rec})}{(1+\varepsilon)^{3/2}}I_2(k,\theta,\theta_A)\Big\};\tag{18}$$

where

$$I_1(k,\theta,\theta_A)=\sqrt{\frac{2\pi}{k\zeta_n\theta}}\sum_{l=0}^{\infty}\sum_{m=0}^{l}d_{lm}J_{2l+\frac{1}{2}}(k\zeta_n\theta)\,_1F_1(1,l-m+\frac{3}{2},-k^2\zeta_n^2\theta_A^2);\tag{19}$$

$$I_2(k,\theta,\theta_A)=\sqrt{\frac{\pi}{2k\zeta_n\theta}}\sum_{l=0}^{\infty}\sum_{m=0}^{l}C_{lm}J_{2l+\frac{1}{2}}(k\zeta_n\theta)\,_1F_1(1,l-m+\frac{5}{2},-k^2\zeta_n^2\theta_A^2);\tag{20}$$

$$d_{lm}=\frac{(-1)^m(2l+\frac{1}{2})\Gamma(l+\frac{1}{2})\Gamma(2l-m+\frac{1}{2})}{\sqrt{\pi}m!l!\Gamma(l-m+1)\Gamma(l-m+\frac{3}{2})};\quad C_{lm}=\frac{d_{lm}}{l-m+\frac{3}{2}};$$

Here $_1F_1(a,b,x)$ is a degenerated hypergeometric function and $\Gamma(x)$ is the gamma function.

At the limit $\theta_A\to 0$ the correlation function $C(\theta,0)\equiv C(\theta)$ has the following form:

$$C(\theta)=\frac{1}{2\pi^2}\int_0^{\infty}dk\,k^2\Phi(k)\left\{\left[3\varepsilon-\frac{\cos\Psi(kr_{rec})}{4\sqrt{1+\varepsilon}}\right]^2+\frac{\sin^2\Psi(kr_{rec})}{(1+\varepsilon)^{3/2}}\right\}\frac{\sin k\zeta_n\theta}{k\zeta_n\theta}.\tag{21}$$

Further we will use the expression (21) for the analytical analysis of the manifestation of the Sakharov oscillations in the angular distribution of $\Delta T/T$ on the celestial sphere. Being based on (21) we introduce the function

$$C_0(\theta)=\frac{a}{2\pi^2}\int_0^{\infty}dk\,k^2\Phi(k)\frac{\sin k\zeta_n\theta}{k\zeta_n\theta},\tag{22}$$

where $a=9\varepsilon^2+\frac{2+\varepsilon}{2(1+\varepsilon)^{3/2}}$. It is easy to see that the correlation function $C(\theta)$ can be written in the following form:

$$C(\theta)=C_0(\theta)-\frac{3\varepsilon}{a\,{}^4\sqrt{1+\varepsilon}}C_1(\theta)+\frac{\varepsilon}{4(1+\varepsilon)^{3/2}a}C_2(\theta),\tag{23}$$

where

$$C_1(\theta)=\frac{\theta-\theta_R}{\theta}C_0(\theta-\theta_R)+\frac{\theta+\theta_R}{\theta}C_0(\theta+\theta_R),\tag{24}$$

$$C_2(\theta)=\frac{\theta-2\theta_R}{\theta}C_0(\theta-2\theta_R)+\frac{\theta+2\theta_R}{\theta}C_0(\theta+2\theta_R),\tag{25}$$

and $\theta_R=\frac{R}{\zeta_n}$.

Thus, at $\theta_A=0$, the investigation of the general properties of the correlation function $C(\theta)$ is reduced to the investigation of the function $C_0(\theta)$. Taking (22) as a basis we introduce the conception of a dispersion $C_0(\theta)$ and of a correlation angle $\theta_c^2=\left[-\frac{C_0(\theta)}{C''(\theta)}\right]|_{\theta=0}$ determined by the form of the spectrum $\Phi(k)$:

$$C_0(0)=\frac{a}{2\pi^2}\int_0^{\infty}dk\,k^2\Phi(k)\tag{26}$$

$$\theta_c^2 = \frac{1}{3\zeta_n^2} \frac{\int_0^\infty dk \ k^2 \Phi(k)}{\int_0^\infty dk \ k^4 \Phi(k)}.$$

For $\Phi(k) \propto k^{-n} e^{-k^2 l_s^2}$ and $n < 3$ it easy to show that

$$\theta_c^2 \simeq \frac{2}{3(3-n)} \frac{l_s^2}{\zeta_n^2},$$ (27)

and at $n = 3$, one has $\theta_c^2 \propto \frac{l_s^2}{\zeta_n^2} \ln \frac{\zeta_n}{l_s}$. Moreover, for further calculations we need the value of the fourth derivative of $C_0(\theta)$ at $\theta = 0$:

$$\frac{d^4 C_0(\theta)}{d\theta^4}\Big|_{\theta=0} = \frac{C_0(0)}{5} \frac{\Gamma(\frac{7-n}{2})}{\Gamma(\frac{3-n}{2})} \frac{\zeta_n^4}{l_s^4}.$$ (28)

Taking into account (26)-(28), the correlation function at $\theta < \theta_c$ can be represented as a Taylor series:

$$C_0(\theta) = C_0(0) \left[1 - \frac{1}{2} \frac{\theta^2}{\theta_c^2} + \frac{1}{4!} \beta \frac{\theta^4}{\theta_c^4} + \dots \right],$$ (29)

where

$$\beta = \frac{4}{45(3-n)^2} \frac{\Gamma(\frac{7-n}{2})}{\Gamma(\frac{3-n}{2})} \frac{\zeta_n^4}{l_s^4}; \quad n < 3$$

Using (26)-(29) let us consider now an asymptotics of the correlation function $C(\theta)$ at $\theta \to \theta_R$. For the range of angles $\frac{l_s^2}{4R^2} < \theta - \theta_R < \frac{l_s}{R}$ one can introduce into (24) the expression (29) for $C_1(\theta)$, and substitute $\theta = \theta_R$ in $C_2(\theta)$. As a result we get the following expression

$$C(\theta \to \theta_R) \simeq C_0(\theta_R) \left[1 - \frac{\varepsilon}{4a(1+\varepsilon)^{3/2}} \right] - \frac{6\varepsilon}{a \sqrt[4]{1+\varepsilon}} C_0(2\theta_R) +$$

$$\frac{3\varepsilon}{4a(1+\varepsilon)^{3/2}} C_0(3\theta_R) - \frac{3\varepsilon}{a\sqrt[4]{1+\varepsilon}} \frac{\theta_c}{\theta_R} C_0(0) \times \left[1 - \frac{1}{2}x^2 + \frac{\beta x^4}{4!} \right]; \quad x = \frac{\theta - \theta_r}{\theta_c}.$$ (30)

One can conclude from this expression that if a spectrum of initial perturbations provides a correlation function which decreases rapidly with the growth of $\theta \gg \theta_c$ (see for example [25], [28]) then the behavior of $C(\theta \to \theta_R)$ at $C_0(\theta_R) \ll \frac{\theta_c}{\theta_R} C_0(0)$ will be determined mainly by the last term in (30). Thus, at the angular scales $\theta \approx \theta_R$ the Sakharov oscillation lead to appearance of a specific anomaly (a "resonance" type) which has a half-width of order θ_c. The analogous situation takes place at $\theta \to 2\theta_R$. In this case

$$C(\theta \to 2\theta_R) \approx C_0(2\theta_R) - \frac{3\varepsilon}{2a\sqrt[4]{1+\varepsilon}} [C_0(\theta_r) + 3C_0(3\theta_r)] +$$

$$+ \frac{\varepsilon}{2a(1+\varepsilon)^{3/2}} C_0(4\theta_R) + \frac{\varepsilon}{8a(1+\varepsilon)^{3/2}} \frac{\theta_c}{\theta_R} C_0(0)\tilde{x} \left[1 - \frac{1}{2}\tilde{x}^2 + \frac{\beta \tilde{x}^4}{4!} \right];$$ (31)

where $\tilde{x} = \frac{\theta - 2\theta_R}{\theta_c}$. One can see from (31), that the results we got earlier for the range $\theta \approx \theta_R$ are also valid at the neighborhood ($/\theta - 2\theta_R/ \leq \theta_c$). We would like to emphasize that the anomalies at $\theta \approx \theta_R$ and $\theta \approx 2\theta_R$ have the same nature. The reason for both of them is the existence of acoustic modes evolving in the cosmic plasma at the epoch of hydrogen recombination. These acoustic modes left traces in the angular correlation function of the CMBR fluctuations due to Silk [13] and Doppler effects.

It is important to emphasize the following fact. In the "resonance" $\theta \sim \theta_R$ the Silk effect only gives the contribution, while in the "resonance" $\theta \approx 2\theta_R$ both effects (Silk and Doppler) make their contributions. Thus we have a unique possibility to search for these effects, which are important both for the clarifying the character of the initial perturbations and for the investigation of the ionization history of the cosmic plasma.

4 Method of filtering of the Sakharov effect in the anisotropy of CMBR.

In the previous section we have described the evidence for acoustic modes in the anisotropy of CMBR. In [25] we called them long distance correlations (LDC) in $\Delta T/T$ distribution. The LDC are thus a specific signature of the acoustic modes. How to filter the microwave background anisotropies which 1) cancels the regular component of the correlations and 2) emphasizes/preserves the effect of LDC?

We proposed the following method [25]: the observer should measure the difference of intensity of CMBR in l_1 and l_3 directions, as well as l_2 and l_4 directions as shown on Fig.2. We introduce an auxiliary function of two variables θ, d as:

$$G_2(\theta, d, \theta_A) = \frac{\langle [T(l_1) - T(l_3)][T(l_2) - T(l_4)]\rangle}{T^2} =$$

$$= C(\theta + d, \theta_A) + C(\theta - d, \theta_A) - 2C(\theta, \theta_A), \tag{32}$$

where $l_1 \cdot l_4 = l_2 \cdot l_3 = \cos\theta$, $l_1 \cdot l_2 = \cos(\theta + d)$, $l_3 \cdot l_4 = \cos(\theta - d)$, T is the average temperature. The value of θ should vary in the range $\theta_R \leq \theta \leq 2\theta_R$ and $d \leq \theta_c$. The method is valid for both the direct measurement with the right beam experiment as well as the specific data-reduction technique.

One can see from (32) that for $d \ll \theta$ the level of anisotropy is related to the second derivative of the correlation function $C''(\theta)$ via:

$$G_2(\theta, d, \theta_A) \simeq C''(\theta, \theta_A)d^2. \tag{33}$$

We show that the quantity $G_2(\theta, d, \theta_A)$ introduced above allows one to measure in an optimal way the contribution of acoustic modes to Δ_{rms}.

On Fig.3 we plot the result of the numerical calculatons of $G_2(\theta, d, \theta_A)$ (using (21) and (33)) for the Harrison-Zel'dovich spectrum of the initial perturbations of the metric and $\theta_A = 0$. The results of the computations of $G_2(\theta, d, \theta_A)$ in the neighborhood of the extrema $\theta \approx \theta_R$ and $\theta \approx 2\theta_R$ can easily be interpreted using

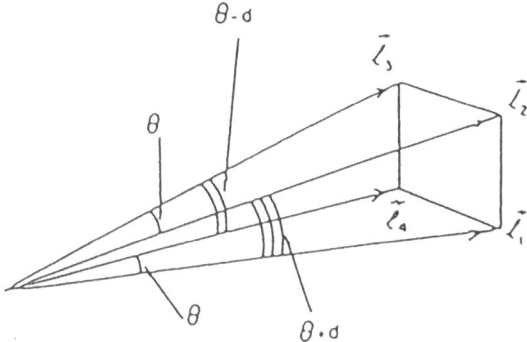

Fig. 2. Space orientation of the unit vectors for the 4-beams experiment.

the properties of $C(\theta)$ which have been found earlier. The LDC effect for the Harrison-Zel'dovich spectrum manifests itself most clearly in the range $\theta \approx \theta_R$ where the last term of (30) dominates:

$$G_2(\theta \approx \theta_R) \approx \frac{9\varepsilon\ d^2}{a\ (1+\varepsilon)^{1/4}} \frac{1}{\theta_R \theta_c} C_0(0)x \left(1 - \frac{5}{18}\beta x^2\right), \qquad (34)$$

where $x = \frac{\theta - \theta_R}{\theta_c}$.

One can see that $G_2(\theta \approx \theta_R)$ is equal to zero at $\theta = \theta_R$ and looks like an almost linear function of x up to the extrema. Further the factor $1 - \frac{5}{18}\beta\ x^2$ plays a significant role (see equation (34) and Fig.3). The analogous phenomenon takes place at $\theta \approx 2\theta_R$, where, however the amplitude of $G_2(\theta \approx 2\theta_R, d)$ is very small (see Fig.3 and eq.(31)).

Let us now consider the dependence of the function $G_2(\theta, d, \theta_A)$ of the value of the half-width of the antenna-beam θ_A. For this purpose we calculated $G_2(\theta, d, \theta_A)$ for $(\theta_A/\theta_R)^2 = 0.1$ (see Fig.3) using the exact definition of $C(\theta, \theta_A)$ (see (18)-(20)). As it is seen on the Fig.3, the peaks are smoothed out for a significant θ_A. Thus for the filtering of the effect under consideration we have to have θ_A, $d < \theta_c$.

Concluding this section we note, that the 4-beam experiment possesses additional possibilities of the investigation of the field of the CMBR. As it follows from Fig.2 in the 4-beam scheme the directions l_1, l_2 and l_3, l_4 form two bi-ray schemes. Using this fact we introduce the following function

$$G_1(\theta, d) = \langle \left[\frac{T(l_1) - T(l_2)}{T}\right]^2 \rangle - \langle \left[\frac{T(l_3) - T(l_4)}{T}\right]^2 \rangle = 2[C(\theta - d) - C(\theta + d)],$$
$$(35)$$

For the angles $\theta \approx \theta_R$ and $\theta \approx 2\theta_R$ at $d \ll \theta_R$ the function $G_1(\theta, d)$ is strictly connected with the value of the first derivative of the correlation function $C(\theta, \theta_A)$ with respect to the angle θ:

$$G_1(\theta, d) \approx -4C'(\theta)d \qquad (36)$$

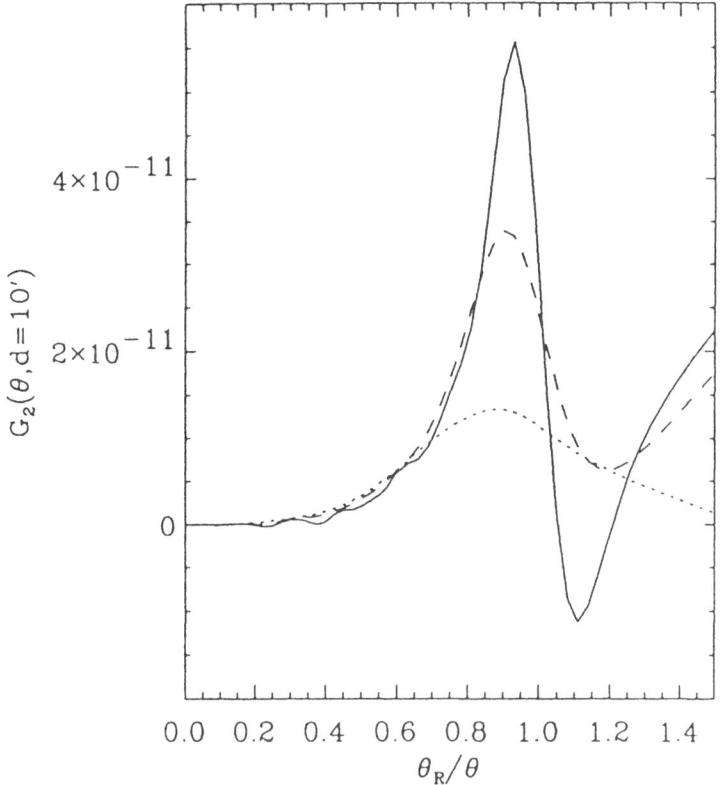

Fig. 3. Function $G_2(\theta, d, \theta_A)$ for the model $\Omega_R = 0.97$; $\Omega_b = 0.03$; $h = 0.5$; $d = 10'$ and the Harrison-Zel'dovich spectrum. Solid line for $\theta_A = 0$, dashed line for $\theta_A = 0.1\theta_R$, pointed line for $\theta_A = 0.3\theta_R$.

where $C'(\theta) \equiv \frac{dC}{d\theta}$.

Using (31) one can evaluate the character of the function $G_1(\theta, d)$ in the neighborhood of the point $\theta \approx \theta_R$. Let us suppose that the main contribution in $G_1(\theta, d)$ is given by the last term of (30). Then

$$G_1(\theta \approx \theta_R, d) \approx \frac{12\varepsilon}{a(1+\varepsilon)^{1/4}} \frac{d}{\theta_R} C_0(0) \left(1 - \frac{3}{2}x^2 + \frac{5}{24}\beta\, x^4\right). \qquad (37)$$

As one can see from (37), that the function $G_1(\theta = \theta_R, d)$ reaches an extremum at the point $\theta = \theta_R$ and has a form of a distinct peak. This result is confirmed by the data of numerical computations of $G_1(\theta, d)$ which are shown on Fig.4. It is seen that in this case the formation of peak-like anomalies takes place at the points $\theta = \theta_R$ and $\theta = 2\theta_R$ where the function $G_2(\theta, d)$ had the asymptotics $G_2(\theta \to \theta_R, d) \propto (\theta - \theta_r)$ and $G_2(\theta \to 2\theta_R, d) \propto (\theta - 2\theta_r)$.

We have to mention another important property of the functions $G_1(\theta, d)$ and $G_2(\theta, d)$. As we has shown above, the half-width of the peaks of the function

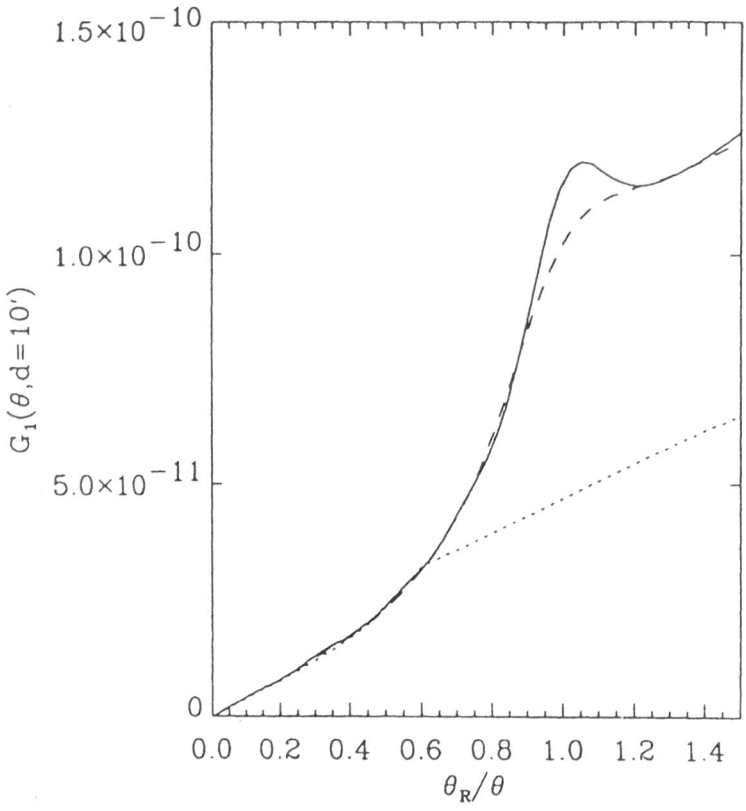

Fig. 4. Function $G_1(\theta, d, \theta_A)$ for the model $\Omega_R = 0.97$; $\Omega_b = 0.03$; $h = 0.5$; $d = 10'$ and the Harrison-Zel'dovich spectrum. Solid line for $\theta_A = 0$, dashed line for $\theta_A = 0.1\theta_R$, pointed line for $\theta_A = 0.3\theta_R$.

$G_2(\theta, d)$ at $\theta \approx \theta_R$ is determined by the coefficient β (see (34)). For the function $G_1(\theta, d)$ the half-width of the peaks is strictly determined by the angle θ_c (see (37)). Thus, using the properties of the functions $G_1(\theta, d)$ and $G_2(\theta, d)$ at $\theta \approx \theta_R$ we can, in principle, get θ_c and β after estimation of the half-width of the peaks. Both these parameters play an important role in the determination of the statistical properties of the random temperature fluctuations of the CMBR on the selestial sphere (Zabotin and Naselsky [35]; Sazhin [36]; Bond and Efstathiou [7]).

5 Sakharov effect in the spatial correlation function of galaxies.

In the previous sections we analyzed the manifestations of the Sakharov oscillations in the distribution of $\Delta T/T$ on the celestial sphere and discussed possible ways of its filtering. The question arises, if it is possible to find observational indications of the Sakharov oscillation in the distribution of the galaxies and clusters of galaxies For the first time this problem was pointed out in the Sakharov's paper [12] and was discussed in the monograph of Zel'dovich and Novikov [15]. We will show that the answer to this question crucially depends on the properties of the dark matter in the Universe and mainly on whether this dark matter is baryonic or non-baryonic matter (Jørgensen et al. [38]).

Let us consider first the standard CDM models. For these models the present density of the non-baryonic dark matter is remarkably larger than the baryonic dark matter density ($\Omega_R \gg \Omega_b$). In this case the acoustic modes in the plasma at the epoch of the hydrogen recombination make relatively small additional perturbations of the gravitational potential which is determined mainly by non-baryonic dark matter:

$$\delta\phi_b \propto \frac{4\pi G\rho_b\delta_b\tilde{R}^2}{k^2}; \ \delta_b \equiv \frac{\delta\rho_b}{\rho_b}.$$

This add to the potential leads to the modulation of the density of non-baryonic dark matter at $z = z_{rec}$ on the level

$$\tilde{\delta}_R = \frac{\delta\tilde{\rho}_R}{\rho_R}|_{z=z_{rec}} \simeq \frac{z_{rec}}{z_{ev}}\frac{\Omega_b}{\Omega_R}\delta_R(z_{rec}) \simeq 3 \cdot 10^{-3}\frac{\Omega_b}{\Omega_R^2}h^{-2}\delta_R(z_{rec}) \qquad (38)$$

where $\delta_R(z_{rec})$ is the "standard" contrast of the density fluctuations in the CDM model without taking into account the influence of inhomogeinities of the baryonic matter.

As it is seen from (38), the level of the spectrum modulations of the cold gravitating particles depends not only on the ratio of Ω_b and Ω_R, but also on the absolute value of Ω_R. In particular, for the standard CDM model with $\Omega_b = 0.03$, $\Omega_R = 0.97$ and $h = 0.5$ the level of modulations of perturbations in the cold particles is $\tilde{\delta}_R/\delta_R(z_{rec}) \simeq 4 \cdot 10^{-4}$ which is negligibly small. However, the level of modulation is on the other hand remarkable $\tilde{\delta}_R/\delta_R(z_{rec}) \simeq 0.12$ for models with $\Omega_b = \Omega_R = 0.1$, cosmological constant $\Omega_\lambda = 1 - \Omega_b - \Omega_R$ and, $h = 0.5$,

This effect also influence the correlation function $\xi(r)$ of the initial density fluctuations (see for example Holtzman and Primack [39]). The correlation function $\xi(r)$ has a weak anomaly around $r \simeq 100$Mpc, which is connected with the effect from the acoustic modes of the baryonic plasma on the density fluctuations in the nonbaryonic dark matter. In the models with $\Omega_b \ll \Omega_R$ this effect becomes vanishing small.

It follows from the above discussion, that the strongest effect of the Sakharov oscillations in the spatial distribution of galaxies could be expected in the case where the dark matter in the Universe is baryonic. In this case the initial inhomogeneities in the plasma can only be of isocurvative type, because the adiabatic modes are incompatible with the data on $\Delta T/T$ and large scale correlations of galaxy distributions. Jørgensen et al. [38] discussed the correlation function of galaxies for the model with initial isocurvature perturbations and showed that for this type of inhomogeneities the peak-like anomalies arise at $r \simeq R$

and $r = 2R$, where R is the acoustic horizon at the epoch of recombination. On Fig.5a,b we have plotted the correlation function $\xi(r)$ for the isocurvature model with $\Omega_b = 0.3$; $h = 1$; $P_0(k) \propto k^{-\gamma}$ see [38]. The initial spectrum of the fluctuations was smoothed by the Gaussian filter with the halfwidth $R_* = 0.7h^{-1}\text{Mpc}$ (Fig.5a) and $R_* = 3.5h^{-1}\text{Mpc}$ (Fig.5b). It is seen on Fig.5a,b that the Sakharov modulation of the spectrum leads to specific fetures in the function $\xi(r)$ at $r \simeq R$ and $r = 2R$. Fig.6a,b shows the spatial correlation function $\xi(r)$ for the model with $\Omega_b = 0.2$ and $h = 1$. Furthermore we would like to emphasize, that our conclusions are valid not only for the open baryonic models but also for the flat model with the non-zero cosmological constant $\Omega_\lambda = 1 - \Omega_b$ see [38]. In this case we have only a small change of the scale of R.

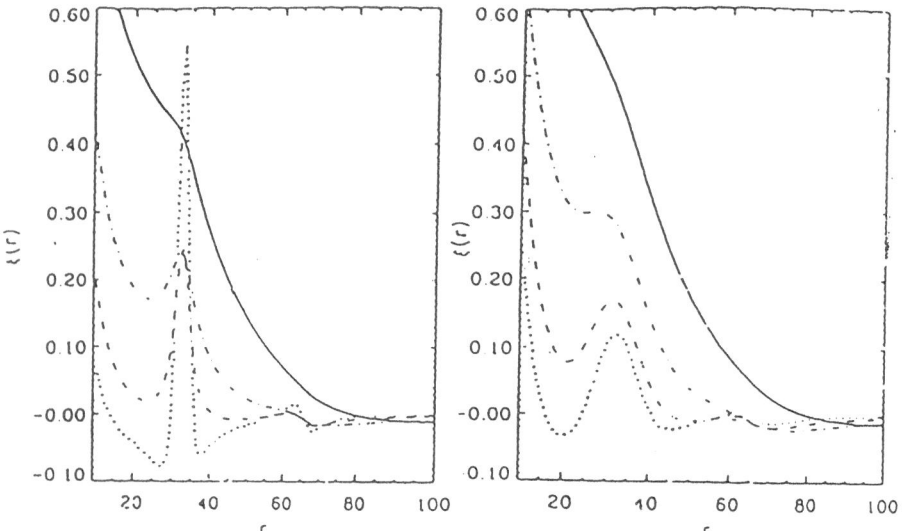

Fig. 5. Correlation function $\xi(r)$ for the model with $\Omega_b = 0.3$, $h = 1$, $P_0(k) \propto k^{-\gamma}$, $\bar{R}_* = 0.7h^{-1}\text{Mpc}$. Solid line corresponds $\gamma = 3$, dahsed line for $\gamma = 2$, dashed-dotted line for $\gamma = 1.2$, doted line for $\gamma = 0.5$. The correlation functions are normalized to 1 at $r = 5h^{-1}\text{Mpc}$. (b) The same as Fig.5a, *but for* $\bar{R}_* = 3.5h^{-1}$ *Mpc*.

What could be the way to filter the correlations in the galaxy distributions?

We propose a method similar to that for the correlations in the anisotropy of CMBR as described in the section 4. It is based on the filtering of the second derivatives of the correlation function. The importance of the determination of the $\frac{d^2\xi}{dr^2}$ was pointed out in [40].

We introduce an auxiliary function of two variables

$$\Phi(r, g) = \xi(r + g) + \xi(r - g) - 2\xi(r). \tag{39}$$

The scale of g should be less than the correlation scale r_c, $g < r_c$. One can see

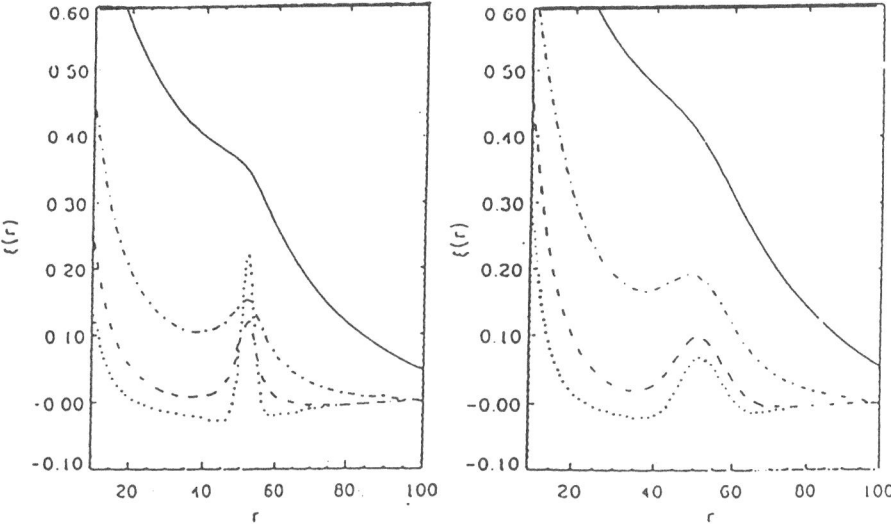

Fig. 6. (a) The same as Fig.5a, but for $\Omega_b = 0.2$. (b) The same as Fig.5b, but for $\Omega_b = 0.2$.

that in the case of $g \ll r_c$ we have

$$\Phi(r, g) = \frac{d^2\xi(r)}{dr^2} g^2. \tag{40}$$

On Fig.7 we plotted $\Phi(r, g)$ around $r = R$ and $r = 2R$ for the baryonic isocurvature model with $\Omega_b = 0.2$; $h = 1$.

6 Conclusion

We have investigated the dynamics of the adiabatic type perturbations at the epoch of recombination using the analytical approach. We have obtained a simple expression for the Fourier components of temperature anisotropies $\left(\frac{\Delta T}{T}\right)_k$ at $z = 0$. We have compared our analytical results with the data of computer modelling of the processes as performed by Muciaccia *et al.* [17] for some specific models (see sec. 2). The dependence of the $\left(\frac{\Delta T}{T}\right)_k$ on Ω_b was also discussed.

We have considered the peculiarities of the angular correlation function for $\frac{\Delta T}{T}$ at the most interesting angular scales $\theta \simeq 20' - 2°$. We will emphasize that the Sakharov oscillations lead (in a CDM model) to an appearance of specific anomalies ("resonances") at the angles $\theta_R = R/\zeta_n$ and $\theta_{2R} = 2\theta_R$.

The dependence of the "resonances" on the parameters of the cosmological model and the width of an antenna was investigated. We pointed out the special importance of the first "resonance" at θ_R for the comparison with observational

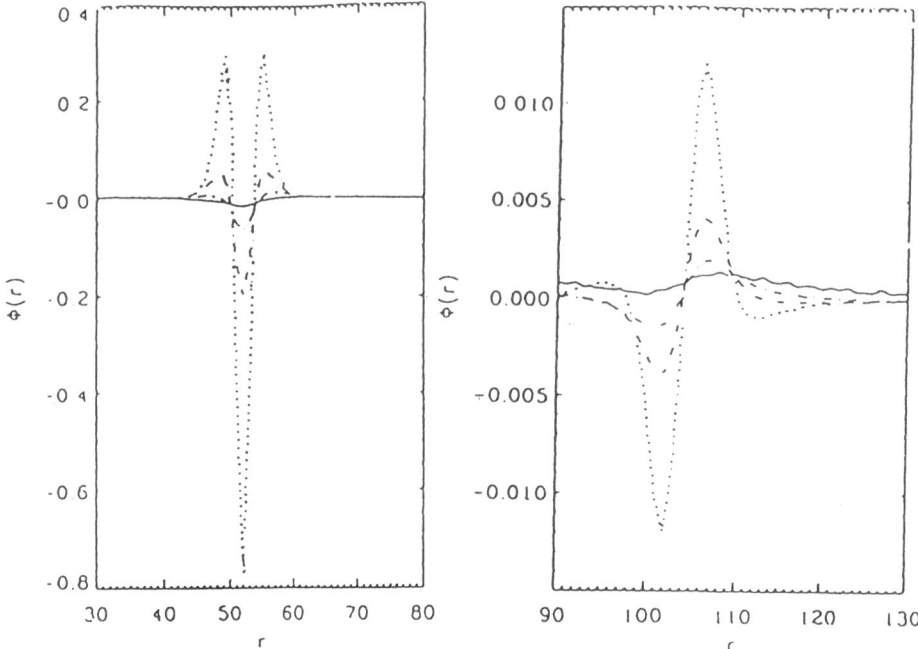

Fig. 7. (a) Function $\Phi(r, d, \theta_A)$ for the model $\Omega_b = 0.2$, $h = 1$ and $d = 3h^{-1}$Mpc at $r \approx R$ for different values of γ Symbols are the same as on Fig.5a. (b) The same as Fig.7a, but for $r \approx 2R$.

data. This first "resonance" is specific for the model with nonbaryonic dark matter and it is absent in the model with baryonic dark matter.

We have proposed a method of filtering of the "resonances" in the correlation function, which is reduced to a direct measure of the second derivative of the correlation function in the vicinity of the "resonance".

The analogous "resonances" appear in the spatial correlation function of galaxy distribution in the models with baryonic dark matter and isocurvature initial perturbations. We also proposed the way to filter the correlations in the galaxy distributions in analogy to that for the correlations in the measurements of the anisotropy of CMBR.

Acknowledgements. This investigation was supported in part by the Danish Natural Science Research Council through grant 11-9640-1 and in part by Danmarks Grundforskningsfond through its support for establishment of the Theoretical Astrophysical Center.

References

[1]   Dressler, A., Faber, S.M., Burstein, D., Davis, R.L., Linden-Bell, D., Terlevich R.Y., and Wegner G. *Ap.J.Lett.* **313** (1987) L37.

[2]   Maddox S.J., G.Efstathiou, W.Y.Sutherland & J.Loveday, 1990. MNRAS **242**, 43.

[3]   Smoot, G.F. *et al. Ap.J.Lett.* **396** (1992) L1.

[4]   Bennett, C.L. *et al. Ap.J.Lett.* **396** (1992) L7.

[5]   Wright, E.L. *et al. Ap.J.Lett.* **396** (1992) L13.

[6]   Tormen, G., Lucchin, F., and Matarrese, S. *Ap.J.* **386** (1992) 1.

[7]   Efstathiou, G., N. Kaiser, W. Saunders, A, Lawrence, M. Rowan-Robinson, R.S. Ellis, and C.S. Frenk: 1990, MNRAS, **247**, 10.

[8]   Saunders, W., C.S. Frenk, M. Rowan-Robinson, G. Efstathiou, A, Lawrence,, N. Kaiser, R.S. Ellis, Y. Crawford, X.-Y. Xia, and I. Parry: 1991, Nature, **349**, 32.

[9]   Vogeley, M.S., C.Park, M.J.Geller & J.P.Huchra, 1992. Ap.J.,**391**, L5.

[10]   Fisher, K.B., M. Davis, M.A. Strauss, A. Yahil, and Y.P. Huchra: 1992, Ap.J. **402**, 42.

[11]   Peacock J.A., 1991. MNRAS, **253**, 1.

[12]   Sakharov, A.D.: 1965, JETP, **49**, 345.

[13]   Silk, J. *Ap.J.* **151** (1968) 459.

[14]   Peebles. P.J.E., Yu I.T. *Ap.J.* **162** (1970) 815.

[15]   Zel'dovich, Ya.B. and Novikov, I.D. *The Structure and Evolution of the Universe.* (The University of Chicago Press.1983) v.2.

[16]   Fukugita, M, Sugiyama N. and Unemura M., *Ap.J.* **358** (1990) 28.

[17]   Muciaccia, P.F., S. Mei, G. de Gaspers, and N. Vittorio: 1993, Ap.J. **410** L61.

[18]   Doroshkevich, A.G., *Sov. Astron. Lett.* **14** (1988) 125.

[19]   Uson, J.M. and Wilkinson, D.T. *Ap.J.Lett.* **277** (1984) L1.

[20]   Readhead, A.C.S., Lawrence, C.R., Meyer, S.T., Sargent, W.L.V., Hardebeck, H.E. and Moffet, A.T. *Ap.J.* **346** (1989) 566.

[21]   Gaier, T. *et al. Ap.J.Lett.* **398** (1992) L1.

[22]   Meyer, S. *et al. Ap.J.Lett.* **370** (1991) L1.

[23]   Meinhold,P. and Lubin, P. *Ap.J.Lett.* **370** (1991) L11.

[24]   Alsop, D.C., *et al. Ap.J.* **395** (1992) 317.

[25]   Naselsky, P.D., and I.D. Novikov: 1993, Ap.J.**413**,14.

[26]   Wilson, M.L., 1983, Ap.J.**273**, 1.

[27]   Vittorio, N., S. Matarrese, and F. Lucchin; 1988, Ap.J., **328**, 69.

[28]   Naselsky, P.D. and Novikov, I.D., 1993, Proc. of Int.Conf (Erice).

[29]   Jones, B.J.T., and Wyse, R.F.G. *Astr. Ap.* **149** (1985) 144.

[30]   Bond, J.R., Efstathiou, G. *Ap.J. Lett.* **285** (1984) L45.

[31]   Naselsky, P.D., Novikov, I.D., Reznicsky, L.I., *Astr.J.(Soviet Astronomy)* **63** (1986) 1057.

[32]   Wilson, M.L., Silk, J. *Ap.J.* **243** (1981) 14.

[33]   Naselsky, P.D., and Novikov, I.D. *Sov. Astron J.* **33** (1989) 350.

[34]   Doroshkevich, A.G., I.D. Novikov and A.G. Polnarev: 1975, Sov. Astron. **21**, 5, 529.

[35]   Zabotin, N. and P. Naselsky; 1985, Sov. Astron **29**, 614.

[36]   Sazhin, M.V.: 1985, MNRAS, **216**, 25.

[37]   Bond Y.R. & G. Efstathiou, 1987. MNRAS, **226**, 655.

[38] Jørgensen,H., E.Kotok, P.Naselsky & I.Novikov, 1993. NORDITA preprint N93/3A, MNRAS (in press).

[39] Holtzman, J.A. and J.R. Primack, 1993 Ap.J., **405**, 428.

[40] Mo, H.Y., Z.G.Deng, X.Y.Xia, P.Schiller & G. Börner, 1992. A.A. **257**, 1.

[41] Gundersen, J.O. *et al*; 1993, Ap.J. **431**, L1.

[42] Peebls. P.Y.E., 1981, Ap.J. **248**, 885.

Constraints on models from POTENT and CMB anisotropies

Uroš Seljak[12] and Edmund Bertschinger[1]

[1] Department of Physics, MIT, Cambridge, MA 02139 USA
[2] Department of Physics, Jadranska 19, 61000 Ljubljana, Slovenia

Abstract A comparison of density fluctuation amplitudes from POTENT and *COBE* can set stringent limits on various cosmological models. We find that for the standard CDM model the two amplitudes agree well with each other, while some alternative models predict an excessive amplitude for POTENT compared with *COBE*. We show how small scale CMB anisotropy measurements can further constrain the models.

1 Introduction

Determining the abundance and distribution of dark matter in the universe remains a major unresolved problem in cosmology. While a given matter content uniquely determines the shape of the power spectrum (up to variations in the primeval shape), the amplitude of mass fluctuations remains unspecified theoretically and its value must be sought from observations. Traditionally, workers tried to estimate both the shape and amplitude of the power spectrum by mapping the distribution of galaxies in a local volume of space. Unfortunately, this mapping is only indirectly related to the density distribution, requiring an additional assumption about how the light (galaxies) traces mass (dark matter). After the *COBE* discovery of primeval fluctuations (Smoot et al. 1992) it became possible to use temperature fluctuations at the surface of last scattering to compute the amplitude of mass fluctuations, σ_8 (the rms relative mass fluctuation in a sphere of radius $8\,h^{-1}$ Mpc, $H_0 = 100$ h km s^{-1} Mpc^{-1}), for any particular model. This allowed researchers to shift their emphasis from the amplitude determination to the study of the spectral shape. Any additional information on a different scale would enable one to discriminate between different models of structure formation. The *COBE* results are able to provide a constraint on the shape of the primordial power spectrum, but not on the matter content of the universe, because on the scales probed by *COBE* different matter contents produce similar temperature fluctuations. Small scale CMB anisotropy experiments may provide additional information of this kind, but have not yet reached the needed sensitivities.

   An alternative method that is directly sensitive to the mass distribution uses the measured peculiar velocities. Most work to date has been based on estimates of the bulk flow averaged over large volumes (Bertschinger et al. 1990; Courteau et al. 1993). This statistic has several limitations. First, the statistical distribution of spectral amplitude estimates based on the bulk flow is broad (χ^2 with only 3 degrees of freedom) and consequently bulk flows can only weakly constrain the models. Second, bulk flows are sensitive to the systematic errors introduced by nonuniform sampling (the sampling gradient bias of Dekel, Bertschinger, & Faber 1990). Finally, the scales contributing to the bulk flow estimates are large ($\geq$ 40–60 h^{-1} Mpc), whereas most of the peculiar velocity measurements come from smaller distances (10–30 h^{-1} Mpc). It should be possible to place additional independent model constraints on the smaller scales alone.

An attempt to estimate the amplitude of mass fluctuations on intermediate scales using POTENT reconstructed density field has recently been carried out by Seljak & Bertschinger (1993b; hereafter SB). By using a large sample of peculiar velocity data they were able to reduce both the statistical and systematic errors over the previous amplitude determinations that were based on bulk flow estimates (Del Grande & Vittorio 1992; Muciaccia et al. 1993; Tormen et al. 1993). Here we briefly present their method and compare their result to the *COBE* normalization for several different models: standard cold dark matter (CDM), CDM plus a cosmological constant(CDM+Λ), and CDM plus massive neutrinos (CDM+HDM), with several different choices for the Hubble constant and the primeval spectral index. In addition, we discuss the possibility of further constraining the models by using the small scale CMB measurements together with the POTENT result. This should enable one not only to determine the cosmological parameters, but also to test the gravitational instability paradigm itself.

2 Amplitude estimation from POTENT

Assuming a potential flow the present velocity field can be extracted from observed radial peculiar velocities of galaxies by integrating along radial rays. This is the essence of the POTENT reconstruction method (Bertschinger & Dekel 1989; Dekel, Bertschinger, & Faber 1990). Furthermore, in linear perturbation theory there is a simple relation between velocity and density fields: $\delta(\mathbf{r}) = -(H_0 f)^{-1} \nabla \cdot \mathbf{v}(\mathbf{r}) \equiv f^{-1} \widetilde{\delta}(\mathbf{r})$, where $\delta(\mathbf{r})$ is the density fluctuation at position $\mathbf{r}$, H_0 is the Hubble constant and f is the the growing mode logarithmic growth rate, well approximated by $f(\Omega_m) = \Omega_m^{0.6}$.

One can define the likelihood function for a set of measurements of $\widetilde{\delta}$ from POTENT as

$$L(\sigma_8) = \frac{1}{\sqrt{(2\pi)^N |M|}} \exp\left[-\frac{1}{2} \sum_{i=1}^{N} \sum_{j=1}^{N} M_{ij}^{-1} (\widetilde{\delta}_i - \langle \widetilde{\delta}_i \rangle)(\widetilde{\delta}_j - \langle \widetilde{\delta}_j \rangle) \right]. \quad (1)$$

M_{ij}^{-1} and M are the inverse and determinant of the correlation matrix $M_{ij} = \langle (\widetilde{\delta}_i - \langle \widetilde{\delta}_i \rangle)(\widetilde{\delta}_j - \langle \widetilde{\delta}_j \rangle) \rangle$. Here $\langle \rangle$ denotes averaging over the random field ensemble (signal) and the distance errors (noise). For a given theoretical model the correlation matrix depends on the parameters of the power spectrum, most notably on its amplitude. An estimate of the amplitude σ_8 is given by the value that maximizes L. This estimate is, in general, biased (in part because of the assumption of normality, but also owing to the presence of sampling gradient and other biases in the $\{\tilde{\delta}_i\}$). One can estimate the statistical bias by Monte Carlo simulations and correct it. One should also worry about the nonlinear and Malmquist bias effects. However, if the smoothing radius is large then these effects are small even if the unsmoothed density fluctuations are large. SB find that for the smoothing radius of $12\,h^{-1}$ Mpc the two effects are at most of order 10% and opposite in sign, so they can be neglected.

For each model considered, SB estimated the maximum-likelihood value of the amplitude $\sigma_{8,v}$. Because they use a sparse sample (particularly at larger distances) and a large smoothing radius, the results are not particularly sensitive to the shape of the power spectrum. An approximate value valid through most of the parameter space is

$$\sigma_{8,v}\,\Omega_m^{0.6} = 1.3^{+0.4}_{-0.3}\ . \tag{2}$$

95% confidence limit intervals give $\sigma_8\,\Omega_m^{0.6} \sim 0.7 - 2.3$. As a possible caveat we mention that this result is in disagreement with the normalization using the masses and abundances of rich clusters as given by Efstathiou, White, & Frenk (1993). They find $\sigma_8 = 0.57\,\Omega_m^{-0.56}$, which is more than the 2σ away from 2, although including all the sources of systematic errors could bring the two results into better agreement.

3 Amplitude estimation from *COBE*

A useful normalization of *COBE* results is given by Q_{rms-PS}, the ensemble-average rms quadrupole amplitude of CMB fluctuations. Its value can be estimated from the published angular correlation function $C(\theta)$ (Smoot et al. 1992; Wright et al. 1992) and depends on the shape of the primordial power spectrum. For the power law spectrum with $P(k) \propto k^n$ we obtained $Q_{rms-PS} = (15.7 \pm 2.6)\exp[0.46(1-n)]\mu K$ (Seljak & Bertschinger 1993a).

To obtain a small scale normalization $\sigma_{8,l=2}$ from *COBE* one needs to relate the ΔT quadrupole with the mass fluctuation power spectrum for a given model. We have calculated the radiation and density power spectra by the coupled integration of linear density perturbations with contributions from baryons, CDM, radiation and neutrinos using the full set of relativistic Einstein, Boltzmann, and fluid equations. This method does not assume the naive Sachs-Wolfe expression and is valid also when the universe is not matter dominated at the time of recombination (as in the case of low h values) or when $\Omega_\Lambda \neq 0$. In all cases, we fix the baryon contribution to $\Omega_B = 0.0125h^{-2}$ as implied by primeval nucleosynthesis (Walker et al. 1991).

The analysis was restricted to the simplest generalizations of standard CDM model that decrease the power on small scales relative to that on large scales and have been proposed recently as viable models of large scale structure. These models reduce $\sigma_{8,l=2}$ compared with the standard CDM. The decrease can be small, as for example in the CDM+HDM model with $\Omega_\nu \sim 0.2$, or large, as for the tilted model with $n \sim 0.7$, despite the fact that these two models give a similar match to the recent determinations of large scale galaxy clustering power (Maddox et al. 1990). This is because the power spectrum for tilted models differs from its CDM counterpart over all scales between the *COBE* scale and the σ_8 scale (3 orders of magnitude) and not just on small scales as do the CDM+HDM power spectra. Another parameter that affects $\sigma_{8,l=2}$ is the Hubble constant h. It affects the redshift of matter-radiation equality and this changes both the density transfer function and the ΔT versus δ relation. The net effect is that $\sigma_{8,l=2}$ increases for larger values of h.

4 Constraints on the models

In this section we compare the two amplitudes $\sigma_{8,v}$ and $\sigma_{8,l=2}$ to constrain some of the proposed cosmological models. Most of the numerical values in this section are taken from SB. The first model we test is the CDM model. The amplitude predicted by the standard CDM model is consistent with the *COBE* normalization over most of the allowed range of h. For example, $h = 0.5$ gives $\sigma_{8,l=2} = 1.05$, whereas $h = 0.75$ gives $\sigma_{8,l=2} = 1.5$, both of which are compatible with the measured $\sigma_{8,v} = 1.3$. Thus the CDM model cannot be ruled out based on the comparison between the velocity data and *COBE*, a conclusion that agrees with previous comparisons based on the bulk flow estimates on somewhat larger scales (Bertschinger et al. 1990; Efstathiou et al. 1992; Courteau et al. 1993).

Three simple extensions of CDM have recently been proposed to boost the power on large scales relative to that on small scales. The first one adds a cosmological constant Λ with $\Omega_m + \Omega_\Lambda = 1$. This model has an additional attractive feature that it may match both the observationally favored high values of h and the timing constraints from the globular cluster ages. The model, however, fails in the *COBE* versus POTENT comparison unless Ω_Λ is small. While decreasing Ω_m increases $\sigma_{8,v}$ relative to CDM (due to the $f(\Omega_m)$ factor), the time derivative of the potential integrated along the line of sight tends to decrease $\sigma_{8,l=2}$ for a given $\Delta T/T$. For $h = 0.8$ and $\Omega_\Lambda = 0.8$, the $f(\Omega_m)$ factor gives $\sigma_{8,v} = 3.0$, whereas $\sigma_{8,l=2} = 0.66$. Decreasing h even further decreases $\sigma_{8,l=2}$. We find that the results constrain the model to $\Omega_\Lambda < 0.6$ ($h = 0.8$) and $\Omega_\Lambda < 0.4$ ($h = 0.5$), at 95% confidence.

A second recently proposed extension of CDM is the mixed dark matter model which has some massive neutrinos added to the cold dark matter. In this model, the growth factor $f(k)$ depends on the wavenumber because free-streaming damps small wavelengths; $f(k)$ ranges between 1 on large scales to $\frac{1}{4}[(1 + 24\Omega_{CDM+B})^{1/2} - 1]$ on small scales (Bond, Efstathiou & Silk 1980; Ma 1993). This is shown in figure 1 for a particular model with $h = 0.5$, $\Omega_{CDM+B} = 0.7$ and $\Omega_\nu = 0.3$, which corresponds to $m_\nu = 7$ eV. The transition between 1 and 0.8 today ($z = 0$) occurs at roughly $k = 1\,\mathrm{Mpc}^{-1}$ and the effect on $\sigma_{8,v}$ because of this is small. For this model one finds $\sigma_{8,l=2} = 0.67$, which is too small compared with 2. Decreasing the Ω_ν contribution or increasing the value of h reduces the discrepancy. For a particular model of $\Omega_\nu = 0.2$

and $h = 0.75$ one finds that the *COBE* and POTENT amplitudes are in agreement, but, of course, other parameter values with smaller Ω_ν and/or larger h give acceptable results as well.

A third way to decrease small scale power relative to large scales is to tilt the primordial power spectrum $P(k) \propto k^n$ by decreasing n below 1. This again decreases the σ_8 normalization from Q_{rms-PS}, although the best fitted value of Q_{rms-PS} increases (Seljak & Bertschinger 1993a); hence

$$\sigma_{8,l=2} = 1.05\,(1 \pm 0.2)e^{-2.48(1-n)}, \tag{3}$$

similar to the expression based on the $10°$ *COBE* normalization (Adams et al. 1993). Low values of n again imply too small a value of $\sigma_{8,l=2}$ compared with 2. Including a possible gravitational wave contribution to the CMB anisotropy further decreases $\sigma_{8,l=2}$, by a factor of $\sqrt{(3-n)/(14-12n)}$ (Lucchin, Matarrese, & Mollerach 1992). With $h = 0.5$ one obtains $n > 0.85$ with no gravitational wave contribution and $n > 0.94$ with gravitational wave contribution, at 95% confidence limit.

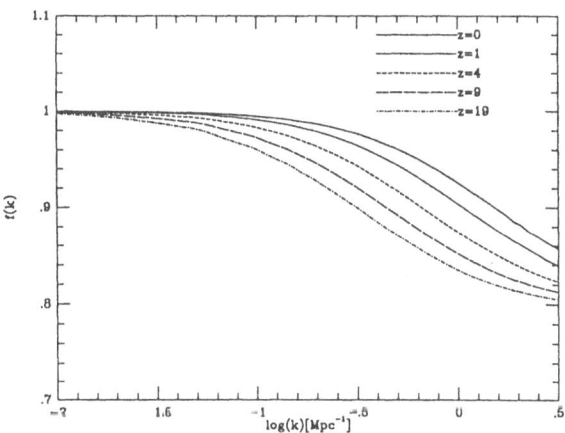

Fig. 1. $f(k)$ versus k at various redshifts for a CDM+HDM model with $h = 0.5$, $\Omega_{CDM+B} = 0.7$ and $\Omega_\nu = 0.3$. For low z the transition occurs at scales smaller than the POTENT smoothing length.

5 $\Delta T/T$ on degree scales

One can also compare large and intermediate scale $\Delta T/T$ measurements or intermediate scale $\Delta T/T$ measurements and POTENT. These comparisons could give additional information on the values of various cosmological parameters. The POTENT scale corresponds to to a CMB angular scale $\sim 0.5°$ so that a direct comparison on those scales would in addition enable one to test the gravitational instability hypothesis itself (Bertschinger, Górski, and Dekel 1990; Górski 1992). Unfortunately, the experimental situation at the moment is still unclear and prevents one to draw any firm conclusions (for a recent review, see Bond 1993), although the experiments currently favor low values of $\Delta T/T$. For this reason we are only presenting theoretical predictions of $\Delta T/T$ for the models discussed in the previous section and we do not attempt to derive any additional constraints from the current observations. Rather, we want to show what parameters could be determined using these types of measurements.

There are several parameters that can affect the $\Delta T/T$ predictions. Figure 2 shows photon transfer functions $l^2 C_l$ for various models discussed in the previous section. We define these by the relations

$$C(\theta) = \left\langle \frac{\Delta T}{T}(\mathbf{n_1}) \frac{\Delta T}{T}(\mathbf{n_2}) \right\rangle = \sum_{l=2}^{\infty} C_l \, P_l(\cos\theta) \ ,$$

where the average is taken over the whole sky for all pairs of directions separated by angle $\theta = \cos^{-1}(\mathbf{n_1} \cdot \mathbf{n_2})$. We normalize all the transfer functions to a fixed quadrupole value C_2.

As fig. 2 shows, $\Delta T/T$ on degree scales ($l \sim 200$) strongly depends on the value of Ω_B (compare curves 1 and 2). Primordial nucleosynthesis constrains the value $\Omega_B h^2$, so the uncertainty in Ω_B mainly reflects the uncertainty in h. Since the value of h at the present is still controversial one cannot determine Ω_B better than within a factor of 4 and this limits the determination of other cosmological parameters as well.

Among the three extensions of CDM discussed in the previous section two of them significantly change the C_l's relative to the standard CDM. A tilt in the initial power spectrum decreases C_l's relative to the $l = 2$ value. For $n = 0.75$ this changes C_{200} by a factor of 3 relative to the CDM value (curve 3 of fig. 2). Ω_Λ also strongly changes the photon power spectrum. Large Ω_Λ values increase $\Delta T/T$ on degree scales (curve 4 of fig. 2) and this may present another difficulty for these models. The third way of changing CDM model by changing some of the CDM into HDM only weakly affects the photon transfer functions (curve 5 of fig. 2) and one cannot hope to determine the value of Ω_ν using these type of measurements.

6 Conclusions

The POTENT analysis of peculiar velocity measurements gives the amplitude of mass fluctuations $\sigma_{8,\nu} \approx 1.3 \, \Omega_m^{-0.6}$. Comparing this value with the *COBE* normalization, one can constrain different models of structure formation. The results strongly challenge the non-zero cosmological constant models and the tilted models. They also point to a higher value of h or a lower value of Ω_ν than have been standard in the mixed dark matter models. A similar comparison based on degree $\Delta T/T$ scales would give additional information on the values of cosmological parameters and could test the gravitational instability paradigm itself.

Acknowledgements: We thank Avishai Dekel for allowing us the use of the POTENT results. This work was supported by grants NSF AST90-01762 and NASA NAGW-2807.

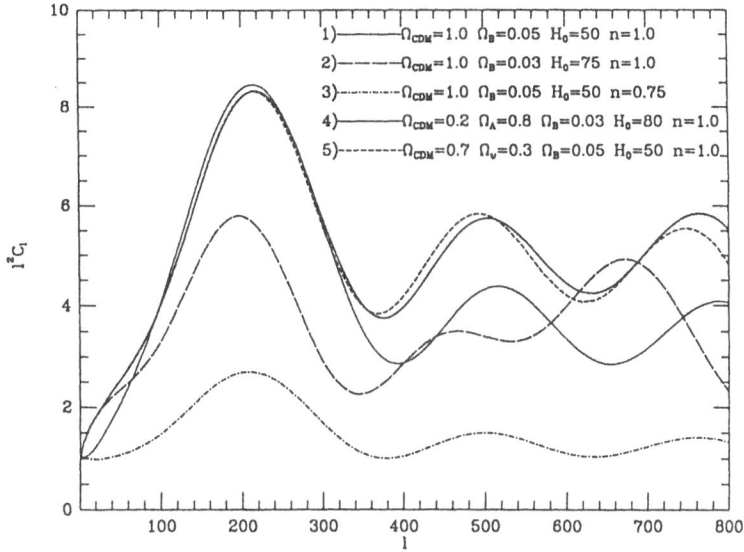

Fig. 2. $l^2 C_l$ for various cosmological models discussed in the text.

References

Adams, F.C., Bond, J.R., Freese, K., Frieman, J.A., & Olinto, A.V. 1993, Phys. Rev. D, 47, 426

Bertschinger, E., & Dekel, A. 1989, ApJ, 336, L5

Bertschinger, E., Górski, K. M., & Dekel, A. 1990, Nature, 345, 507

Bertschinger, E., Dekel, A., Faber, S. M., Dressler, A., & Burstein, D. 1990, ApJ, 364, 370

Bond, J. R. 1993, in Proceedings of IUCAA Dedication Ceremonies, ed. Padmanabhan, T. (New York:Wiley)

Bond, J. R., Efstathiou, G., & Silk, J. 1980, Phys. Rev. Lett. 45, 1980

Courteau, S., Faber, S. M., Dressler, A., & Willick, J. A. 1993, ApJ, 412, L51

Dekel, A., Bertschinger, E. & Faber, S. M. 1990, ApJ, 364, 349

Del Grande, P., & Vittorio, N. 1992, ApJ, 397, 26

Efstathiou, G., Bond, J. R., & White, S. D. M. 1992, MNRAS, 258, 1p

Górski, K. 1992, ApJ, 398, L5

Górski, K., Silk, J., & Vittorio, N. 1992, Phys. Rev. Lett., 68, 733

Lucchin, F., Matarrese, S., & Mollerach, S. 1992, ApJ, 401, L49

Ma, C.-P. 1993, MIT Ph.D. Thesis

Muciaccia, P. F., Mei, S., de Gasperis, G., & Vittorio, N. 1993, ApJ, 410, L61

Seljak, U., & Bertschinger, E. 1993, ApJ (Letters), in press

Seljak, U., & Bertschinger, E. 1993, submitted to ApJ(SB)

Smoot, G. F. et al. 1992, ApJ, 396, L1

Tormen, G., Moscardini, L., Lucchin, F., & Matarrese, S. 1993, ApJ, 411, 16

Walker, T. P., Steigman, G., Schramm, D. N., Olive, K. A., & Kang, H. 1991, ApJ, 376, 51

White, S. D. M., Efstathiou, G., & Frenk C. S. 1993, MNRAS, 262, 1023

Wright, E. L. et al. 1992, ApJ, 396, L13

REIONIZATION AND THE COSMIC MICROWAVE BACKGROUND

Joseph Silk

Departments of Astronomy and Physics, and Center for Particle Astrophysics, University of California, Berkeley, CA 94720, USA

Abstract

If the COBE detection of CMB fluctuations is used to normalize the power spectrum of primordial density fluctuations in a cold dark matter–dominated universe, early reionization is likely to result in a substantial diminution of primordial temperature fluctuations on degree scales. I argue that the reionization may be non–Gaussian because of feedback effects. Secondary fluctuations on arc–minute scales provide an important probe of the efficiency of rescattering.

1 Introduction

We are rightly celebrating at this workshop the triumph of the COBE DMR experiment in measuring fluctuations on $\gtrsim 10°$ angular scales in the cosmic microwave background. This provides the long–sought normalization for the primordial density fluctuation spectrum that gave rise via gravitational instabilities to the present–day structure. Yet there are many choices that remain in models for primordial fluctuations as well as in cosmological parameters. An angular scale of 10° on the last scattering surface corresponds to a comoving scale of 1 Gpc, useful for establishing the validity of the cosmological principle, and for probing inflation, but of little direct relevance to formation of known structure. Only by detecting fluctuations on the precursor scales of the largest structures (degree scales) and galaxy clusters (arc–minute scales) can we hope to specify more precisely the elusive models that quantify the details of galaxy formation. One important complication arises from the thermal history of the universe since redshift $z = 1000$, where the last scattering surface is located in the canonical uniform model of the expanding universe.

The principle purpose of this talk is to argue for the inevitability of reionization in the standard model for large-scale structure, that of cold dark matter, and to discuss, more generally, the secondary signatures of early reionization for microwave background anisotropies on small angular scales. First, however, I describe a new presentation of data from experiments on different angular scales that largely circumvents any assumptions about the radiation power spectrum.

2 Comparison of Different Experiments

Microwave background experiments probe a complex domain of sky and frequency sampling. To facilitate comparison between different experiments, a novel transformation has been developed that transforms the Gaussian autocorrelation function angular fits into something more physically useful.

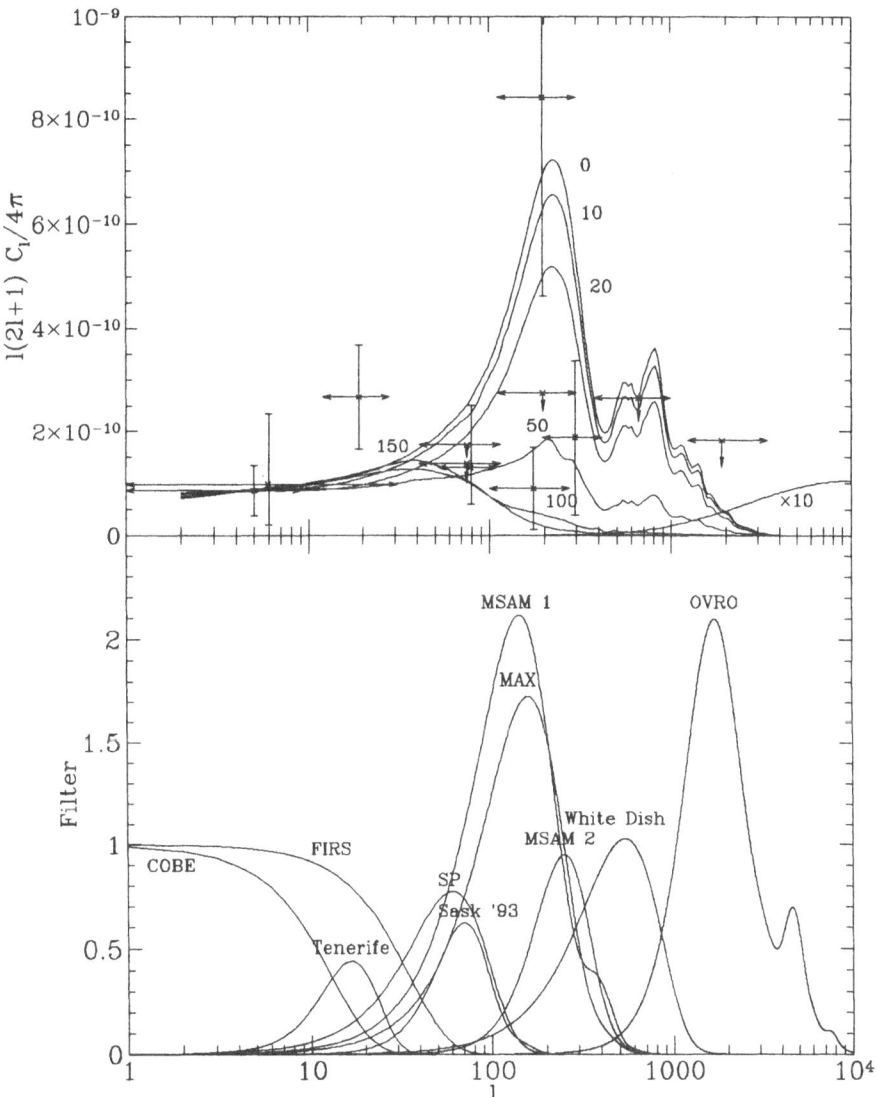

Figure 1. Experimental filter-averaged detections and upper limits, compared to radiation power spectra for CDM $\Omega_B = 0.12$ model for various redshifts of reionization, as labelled (top); experimental filters (bottom)

The approach, pioneered by Bunn [1] and performed independently by Bond [2], is straightforward. The temperature fluctuations are expanded in spherical harmonics $\frac{\delta T}{T} = \Sigma a_{lm} Y_{lm}$ with power per harmonic $C_l = <|A_{lm}|^2>$. The contribution to the total power per decade is $B_l = l(2l + 1)\frac{C_l}{4\pi}$. If the filter function of a particular experiment that defines its sensitivity to harmonics l is F_l, then one can define $\bar{B} = \frac{\Sigma B_l F_l}{\Sigma F_l}$ as the experimental filter function–averaged power. One can show by computing $\bar{B}$ for different input power spectra that it is insensitive to the assumed power spectrum.

Figure 1 displays a comparison between the experimentally-averaged power for various recent detections of cosmic microwave background fluctuations and the radiation power spectra for various cold dark matter models, characterized, in the spirit of the present discussion, by redshifts of reionization that range from 0 to 150 [3]. It is premature to conclude that the apparent disagreement between experiments on degree angular scales is significant, because the uncertainties in modelling out Galactic foreground contamination have not been fully included, and also because of the very limited sky coverage. I discuss below a possible cosmological source of such variations.

3 Reionization with Cold Dark Matter

Early nonlinearity will result in production of dust and ionization input that can reionize the intergalactic medium. Indeed, this must have happened with high efficiency prior to redshift $z \approx 5$. The lack of a Gunn–Peterson absorption trough towards high redshift quasars means that the neutral fraction in the intergalactic medium is less than 10^{-6}. Indirect pointers to early energy input come from the enrichment of the intracluster medium, where there is more iron in the gas than in the stellar components of the cluster galaxies. This requires considerable ejection of enriched gas during the early stages of cluster galaxy evolution. Ejection from supernovae would inevitably involve input of heat into the intergalactic medium. Another indirect indicator of early heat and ionization input comes from the excess, relative to the present day, population of faint blue dwarf galaxies required to account for the deep galaxy counts. The lack of a present day counterpart in the field requires such galaxies, also a common feature in theoretical models, to have ejected considerable amounts of their initial mass in a supernova–driven wind. In nearby galaxy clusters, there is a population of low surface brightness dwarfs that may be the relics of these early, star forming dwarfs: again, considerable mass loss must presumably have occurred in order to differentiate the low surface brightness dwarfs from the compact dwarf galaxies.

If there is so much energy input at $z \lesssim 5$, it is not much of an extrapolation to imagine that a more modest energy input occurred at an earlier epoch. Only ~ 20eV per baryon, a miniscule fraction of the proton rest mass, is required to photoionize the intergalactic medium. Once ionized, the gas can recombine. However the ratio of recombination time to expansion time is approximately $0.06h^3(100/(1+z))^{\frac{3}{2}}$. Hence the required efficiency of conversion of mass into, say, ionizing photons is still low even at $z \sim 100$.

Various structure formation models are capable of early ($z \gtrsim 50$) reionization. These include models with intrinsically nonlinear objects, such as texture or string–seeded models. Another class of models in which early reionization is natural and even inevitable is the baryon dark matter–dominated universe. These models are only viable if the primordial density fluctuations are isocurvature. The spectrum of isocurvature density fluctuations imposed by large–scale structure constraints [4] $\frac{\delta\rho}{\rho} \propto M^{-\frac{n+3}{6}}$, $n \approx -0.5$, diverges towards small scales; $n \lesssim 0$ is required observationally to avoid excessive power on Mpc scales, and $n < 1$ guarantees divergence in terms of potential fluctuations on small scales early in the matter-dominated era. These models can even plausibly bypass a neutral period: there need be no hydrogen recombination epoch. The more conventional, cold dark matter–dominated universe, if normalized to reproduce the COBE DMR measurement of $\frac{\delta T}{T} \approx 10^{-5}$ on large angular scales, has early collapse of small, rare objects. These can, with reasonable efficiency, reionize as early as $z \sim 100$.

Many models fail to have sufficient power to reionize at an epoch sufficiently early that some smoothing of CMB fluctuations occurred. These include mixed dark matter (30 percent hot, 70 percent), tilted ($n \approx 0.7$), and vacuum-dominated cold dark matter ($\Omega_{vac} \approx 0.8$) models, all designed to reconcile COBE $\frac{\delta T}{T}$ with large–scale ($\sim 1 - 50h^{-1}$Mpc) structure observations. In a cold dark matter–dominated universe, however, with mass tracing light and $\Omega = 1$, the typical object to go nonlinear at $z \sim 30$ has mass $\sim 10^4 M_\odot$.

An investigation of reionization in a cold dark matter–dominated universe led to the following conclusions [5]. The fraction of the rest mass density of nonlinear objects that goes into ionizing photons is taken to be a free parameter. With one percent efficiency of producing ionizing photons and unbiased cold dark matter, ionization is complete at $z = 80$. If $f = 0.006$, as might be appropriate if 90 percent of the baryons formed massive stars that produced ionizing photons, reionization occurs as early as $z = 130$. It is difficult to imagine more efficient conversion of baryons into ionizing photons.By contrast, a tilted model could not reionize before $z \sim 20$.

If the universe is reionized back to redshift z, the scattering probability of CMB photons depends only on the combination $\Omega_b h$. From primordial nucleosynthesis, $\Omega_b h \approx 0.015$, so that for a cold dark matter universe, one might take $\Omega_b h \approx 0.03$ and infer a scattering probability of 0.7 at $z = 100$, 0.4 at $z = 60$, 0.2 at $z = 30$. The corresponding angular scales subtended at these epochs are $4°, 5°$, and $6°$. For $\Omega h = 0.03$, the probability distribution for the redshift which a CMB photon was last scattered, or visibility function, peaks at $z \approx 40$. For a CMB photon to have been scattered at least once with probability 0.5 only requires an efficiency $\sim 10^{-4}$ in a canonical model after typical low mass objects undergo first collapse at $z \sim 40$.

4 Generation of Secondary Fluctuations

Reionization at $z \gtrsim 20$ partially smooths out degree-scale CMB fluctuations, but regenerates small angular scale fluctuations. This second order effect arises because in first order, no fluctuations arise. To see this, one studies the photon collision terms that are source terms in the second order Boltzman equation for the radiation intensity, Δ (equivalent to $\frac{\delta \rho_\gamma}{\rho_\gamma}$), namely

$$\frac{\partial \Delta}{\partial t} + \gamma_i \frac{\partial \Delta}{\partial x_i} = n_e \sigma_T [1 + \delta(\mathbf{x})] \left\{ -\Delta + 4\gamma.\mathbf{v} - v^2 + \ldots \right\}.$$

Here γ_i is the photon direction, $\delta(\mathbf{x})$ is the first order matter fluctuation at position $\mathbf{x}$, $\mathbf{v}$ is the matter velocity. To first order, the right hand side of the Boltzman equation has a term $(-\Delta n_e \sigma_T)$ that, arising from primary last scattering, is exponentially suppressed by rescattering as $\exp\left(-\int n_e \sigma_T c \, dt\right)$. The Doppler term linear in v, $\sim n_e \sigma_T v$, is suppressed by phase cancellation. The second order terms survive. However the $O(v^2)$ and other terms are negligible, and the only significant contribution term is the $O(\delta v)$ term. This is known as the Vishniac effect. Since it is in the convolution of two terms that, in the BDM model, at most diverge weakly towards smaller scales when averaged over the thickness of the last scattering surface $\delta \propto R^{-\frac{n+3}{2}}$, $v \propto R^{-\frac{n+1}{2}}$, it diverges significantly, $\frac{\delta T}{T} \propto \delta v \propto R^{-n-2}$, $n \approx -\frac{1}{2}$, and is of order 10^{-5} on subarcminute scales if $\Omega \sim 0.2$. In reionized CDM models, the Vishniac effect amounts at most to $\frac{\delta T}{T} \sim 10^{-6}$ for any realistic ionization history and baryon density.

Since observational limits are about $\frac{\delta T}{T} \sim 10^{-5}$ over an arcminute, from experiments at OVRO and ATCA, one can strongly constrain the allowable (Ω, n) parameter space for the BDM model [6]. Some parameter space remains, near $n \approx -0.5$, $\Omega \approx 0.1 - 0.2$, where the BDM model happens to have the parameters that are preferred by large–scale structure observations.

5 Reionized CDM

While the CDM predictions on small scales are well below current limits, the factor $\lesssim 2$ suppression that is likely on degree scales is significant. If one takes the lowest limits to date on degree scales at face value, then consistency with the COBE detection on $10°$ and the CDM prediction for $n = 1$ initial conditions has been questioned. Smoothing makes a significant difference to this conclusion: even reionization in a CDM model at $z = 20$ suffices to remove any discrepancy between the nucleosynthesis range for Ω_b and $n = 1$ COBE–normalized CDM [3].

Early reionization results in another possibility that has interesting observational consequences. Since the reionization is a consequence of the early non–linearity of very rare dwarf galaxies, one can well imagine that there may be non–Gaussian smoothing of the primordial Gaussian fluctuations. This could arise as follows. The first objects to form, in the extreme tail of the distribution, ionize their environments. Suppose that this local reionization induces a feedback effect. For example, reionization can actually *enhance*, (*e.g.* [7]) the formation of H_2 molecules that form via H^- ions. The enhanced cooling and gas dissipation would enhance dwarf galaxy formation with this positive feedback. One can imagine this process bootstrapping, much as in the early models of explosive galaxy formation [8, 9] that are now ruled out by the low limits on the Compton y−parameter [10]. With partial reionization, however, being the contagious carrier that induces galaxy formation rather than explosion-generated overpressure, one can avoid a hot intergalactic medium, yet still propagate the seed-induced effect out to an appreciable fraction of the horizon at last scattering. Hence the new last scattering surface and reionization would not occurr simultaneously in cosmic time. The reionization process would be patchy and non−Gaussian. This could well result in variable optical depth or thickness of the last scattering surface over degree scales.

6 Summary

There is no doubt that early reionization is an essential feature of the observed Universe. It certainly occurred by $z \sim 5$, and most likely occurred before $z \sim$ 20 in unbiased CDM models. Diminution of $\frac{\delta T}{T}$ on degree scales by a factor $\lesssim 2$ is almost inevitable in COBE−normalized CDM. Surviving fluctuations are plausibly non−Gaussian, since the reionization physics of rare objects could well introduce local feedback. A consequence is that reionization and rescattering generate potentially observable secondary fluctuations on arc−minute scales. In CDM models, these secondary fluctuations are $\frac{\delta T}{T} \sim 10^{-6}$. However in BDM models, they must be of order 10^{-5}, and already provide a strong constraint on the allowable BDM parameter space.

I thank E. Bunn, W. Hu, D. Scott, N. Sugiyama and M. Tegmark for valuable discussions.

References

1. Bunn, E., in preparation (1993)
2. Bond, J. R., in press (1993)
3. Sugiyama, N., Silk, J. & Vittorio, N., Ap. J.Letters, in press (1993)
4. Cen, R. Y., Ostriker, J. P. and Peebles, P. J. E., preprint (1993)
5. Tegmark, M., Silk, J. and Blanchard, A., Ap. J., in press (1993)
6. Hu, W., Scott, D. and Silk, J., Ap. J., in press (1993)
7. Kang, H. and Shapiro, P. R., Ap. J., 386, 432 (1992)
8. Ostriker, J. P. and Cowie, L. L., Ap. J., 243, L127 (1981)
9. Ikeuchi, S., Pub. Astr. Soc. Jap., 33, 211 (1981)
10. Wright, E. L. et al., Ap. J., in press (1993)

Possible Reionization and First Structures in CDM

Alain Blanchard

Université Louis Pasteur, Observatoire astronomique, 11, rue de l'université
67 000, Strasbourg, France

1 Introduction

In the Cold Dark Matter theory, structure formation is known to occur lately.
This is specially true when one is dealing with galaxy formation in $b \approx 2$ normalization, for which the redshift formation could be as late as $z \approx 1$. However,
for a higher normalization, as required by COBE ($b \approx 1$), the formation of first
structures occurs quite early. For instance, Blanchard et al. (1992) have shown,
that at least 10% of the cosmological gas could have been cooled by a redshift
10. Tegmark et al. (1993) showed that in the ($\Omega = 1$, $b = 1$, $h = 0.5$) CDM
model a partial reionization can occur at redshift ≈ 50 by objects with mass of
the ordre of $10^6 M_\odot$, which may partly smeared out the fluctuations on small
anguler scales. In the present paper, I would like to point out that the very first
structures actually form early, with masses of a few $100 M_\odot$ and virial temperature of the order of 100 K. Such objects are likely to form VMO's and migh
allow reionization to occur early enough to fully smear out these fluctuations.

2 First Structures

In a hierarchical picture, fluctuations in the matter field exist on every scales,
and increases indefinitly as the scale is smaller :

$$\sigma(m) \propto m^{-\alpha} \tag{1}$$

because of this there are no "first structures" to form for the dark matter. However, these structures do not present any interest if baryons can not fall inside.
This situation will certainly occur if the temperature of the matter is larger
than the virial temperature of the collapsing object. The virial temperature of a
collapsing mass i.e. the temperature of the gas in hydrostatic equilibrium in the
potential is:

$$T = 50 \text{ K } \left(\frac{1+Z}{100}\right)\left(\frac{M}{10^3 M_\odot}\right) \tag{2}$$

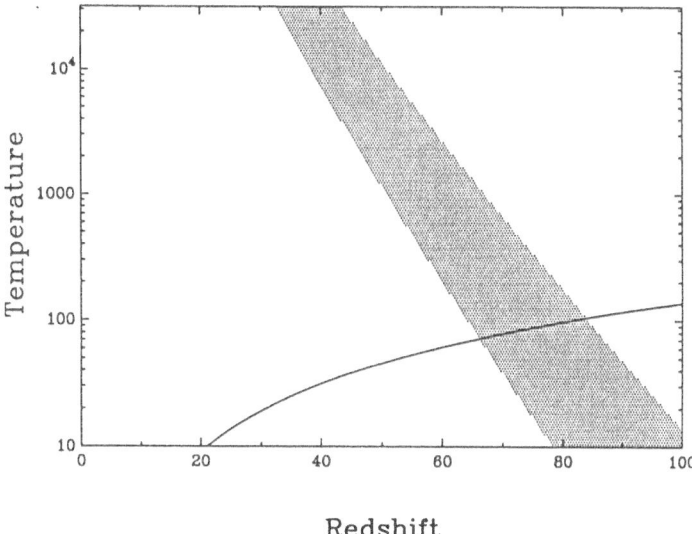

Redshift

Fig. 1. The grey area represents the caracteristics (temperature) of first structures $(3.5 - 4.5\sigma)$ versus redshift. The solid line represents the matter temperature. This graph demonstrates that no structures (i.e. less than 10^{-4}) can form before $Z \approx 85$ while at $Z \approx 60$

substantial fraction of objects have form with masses in the range 3 000 to 500 000$M_\odot$. The associated objects have a baryonic mass between 150 to 25 000$M_\odot$ with virial temperature between 60 and 2000 K.

See Blanchard et al. (1992). Fluctuations above 4.5 σ contain 10^{-5} to 10^{-4} of the total mass (the first value is estimated from Press and Schechter, which is likely to underestimate the actual fraction; the latter value is estimated from the peak formalism which is more likely to give a better value for rare fluctuations). In the following we argue that fluctuations that may allow a full reionization are 3.5–4.5σ. Consequently, we can compute the temperature versus redshift of these fluctuations when they are collapsing. This range is presented on Fig. 1 by the grey area. The temperature of matter is given by the full line. The first structures to form, with baryons falling in, correspond to the intersectioni, occuring at redshift $\approx$ 60–80 with typical temperature 60–100 K. The mass range is 2800–3600 M$_\odot$. The baryonic mass is therefore 150–300 M$_\odot$. Because of the very low temperature, such object are likely to form stars rapidly.

3 Possible Reionization

Very massive stars can be very efficient in nuclear burning. Bond et al. (1984) did estimate that such object can convert 5.10^{-3} of their rest mass in radiation, which can easily ionize quickly the intergalactic medium. Moreover the remnant of such objects are likely to be black holes. Latter accretion on the black hole migh release a considerable ammount of energy (more than 100% of the rest mass actually!). At high redshift the Compton cooling is extremely efficient, so in order to maintain the ionization one needs 1000 eV per baryon. The energy available per baryon from the ionising sources is:

$$E \approx F_c m_p c^2 \epsilon \tag{3}$$

where F_c is the collapsed fraction in the reionising objects, m_p is the proton mass and ϵ is the efficiency of the source. With $\epsilon \approx 5.\,10^{-3}$, the collapsed fraction which is required is $2.\,10^{-4}$. The conclusion that we can draw are very unsensitive to the exact asumption about the required fraction F_c of objects satisfying our criteria $(T_v > T_m)$: at $z = 100$ this fraction is 10^{-9}, while $z = 50$ it is 10^{-2}.

4 Conclusions

The formation of the first structures will be determined by the epoch at which the virial temperature of the collapsing structures will be larger than the temperatre of the cosmological gas. In CDM theory, normalized to COBE data, this occurs at redshift of the order 60–80, allowing for a full reionization.

References

[1992] Blanchard, A., Valls-Gabaud, D., Mamon, G.: Astron. Astrophys. **264** (1992) 365–378

[1982] Bond, J.R., Carr, B.J., Arnett, W.D.: Astrophys. Journal **280** (1984) 809–918

[1993] Tegmark, M., Blanchard, A., Silk, J.: Astrophys. Journal *in press.*

CMB Anisotropies in the Reionized Universe

Naoshi Sugiyama

[1] Astronomy Department, Campbell Hall, University of California, Berkeley, CA94720, USA***

[2] Department of Physics, Faculty of Science, University of Tokyo, Hongo, Bunkyo-ku, Tokyo, 113, Japan

1 Introduction

Observations of Cosmic Microwave Background (CMB) anisotropies smaller than the 10 degree scale are more and more important after the discovery of cosmic microwave background (CMB) anisotropies by COBE satellite (Smoot et al. 1992). If we assume the standard thermal history, one degree scale is just correspond to so-called doppler peak on CMB power spectrum because it is the horizon scale of the last scattering surface, i.e, the horizon scale at the standard recombination. Supposing possible reionization after the epoch of standard hydrogen recombination, however, we find the place of last scattering becomes much closer to present and the horizon scale increases. Hence, this peak is very sensitive to the thermal history of the universe.

Readers will find the motivation, the physical mechanism and the constraint from compton y-parameter of reionization after the standard recombination in a paper by J. Silk of this proceeding. Here, we would like to point out the fact that reionization reduces CMB anisotropies. Recently, several authors suggest cold dark matter dominated models (CDM models) are no more preferable (Gorski et al. 1993) or are given very severe constraints (Dodelson and Jubas 1993) from the resent observations by Gaier et al. (1992) at South Pole(hereafter as SP91). The beam of this experiment had a sinusoidal throw on the sky with amplitude $1.°5$. Though the doppler peak of the CDM model without reionization is slightly smaller than this scale, this experiment is still sensitive to this peak. If reionization occurs, however, such a high peak is smeared out and CDM models may be saved.

There is another model motivated by reionization. From the numerical simulations by several authors (Suginohara and Suto 1992 and Cen, Ostriker and Peebles 1993), the primeval isocurvature baryonic models (hereafter as PIB models) are known as one of the most attractive candidate of the large scale structure formation. Because PIB model has very steep spectrum, first objects created in the very early stage and the universe kept ionizing till present. The most preferable spectrum index n from numerical simulations is about -0.5. Then when we

---

*** e-mail: sugiyama@bkyast.berkeley.edu

consider PIB models, we naturally assume reionization in the very early stage just after the recombination.

In this paper, we will show how reionization after the recombination affects the CMB anisotropies. And constraints on several specific models, i.e., CDM models and PIB models are given from recent observations.

2 Behaviour of CMB anisotropies in the Reionized Universe

In oder to know the behaviour of CMB anisotropies through reionization, let's consider the *standard* CDM model, i.e., with density parameter $\Omega = 1$, Harrison-Zel'dovich initial spectrum, initially adiabatic perturbations as an example. The perturbation equations are soleved numerically in the gauge invariant method (see e.g. Sugiyama and Gouda 1992). The features of CMB spectrum of standard CDM models without reionization are well known as; on the large scale, the Sachs-Wolfe effect is dominated and this plateau on CMB spectrum are shown in Fig.1(a) (dashed line). There is a doppler peak corresponds to the horizon scale of the last scattering surface on the intermediate scale. If reionization did not occur or occurred at very recent epoch, scattering between photons and electrons is not effective after the recombination epoch and the universe is transparent. Hence the scale of the last scattering surface is the horizon scale at the recombination epoch. On small scales, you can find significant damping because of photo diffusion (Silk 1968). This scale corresponds to the thickness of last scattering surface.

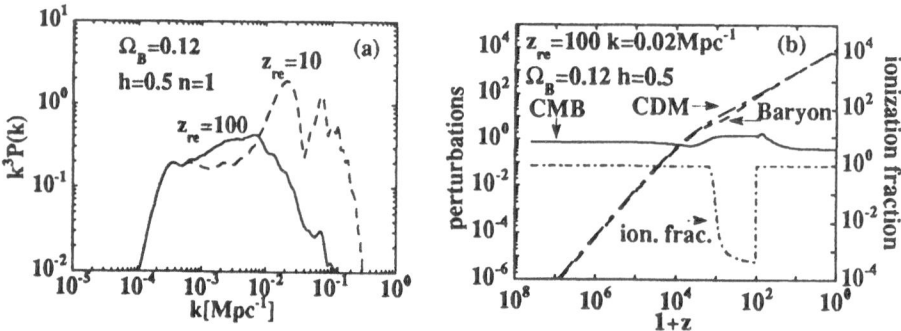

Fig. 1. (a)Power spectrum of temperature anisotropies for CDM models. Dashed line and solid line represent models with reionization at $z = 10$ and $z = 100$, respectively. The normalization is arbitrary. (b)Time evolution of temperature anisotropies, CDM and Baryon density fluctuations as a function of red-shift. The ionization fraction is also shown.

What will happen if reionization occurs? In Fig.1(b), the time evolution of perturbations in the reionized universe is shown. The scale of perturbations coincides with the original doppler peak. It is found that the perturbation of temperature anisotropies grows after entering the Jeans scale. Once recombination occurs, however, this perturbation remains constant because of the free streaming behaviour. Then the perturbation is suddenly damped if we assume reionization at $z = 100$ as shown in this figure. The perturbations on the scale which is smaller than the horizon scale of the new last scattering surface are smeared out. This new last scattering surface stays where the optical depth equals unity roughly written as $z = 65\Omega(0.05/\Omega_B h)^{2/3}$. Here Ω, Ω_B and h are present total density parameter, present baryon density parameter and Hubble constant normalized by 100km/s/Mpc, respectively. The resultant spectrum is also shown in Fig.1(a) (solid line). You can find the extreme suppression of the doppler peak. On the other hand, there exists a newly created peak on the large scale. This scale corresponds to the horizon scale at the new last scattering surface. Here we have to mention that the effect of second oder perturbations is important especially for PIB models because the last scattering surface is so close to present. On small scale temperature anisotropies, this effect, that is, Vishniac effect (Vishniac and Ostriker 1986 and Vishniac 1987) is dominated for PIB models.

In Fig.2, in oder to directly compare the spectrum with specific observations, we show the coefficients of the temperature anisotropies C_ℓ in ℓ space as a parameter of the epoch of reionization for a CDM model (a). Evidently, reionization is effective at significantly reducing the Doppler peak. The same power spectrum for the PIB model is shown in Fig.2 (b). The contribution of the Vishniac effect is also shown on these figures. And in Fig.2(c), the window functions W_ℓ of SP91, OVRO(Readhead et al. 1989) and COBE are shown. The expected temperature anisotropy of each experiment is expressed as $(\delta T/T)^2 = \sum_{\ell>1}(2\ell+1)C_\ell W_\ell/4\pi$.

3 Constraints on Models

The method to get the constraints on the specific model is as follows. We use the observed value 30μK by COBE as the normalization of CMB anisotropies. Then we check the value of σ_8 and compare the expected anisotropies with OVRO experiment, SP91 and the quadrupole anisotropies by COBE. In oder to treat SP91, we use the Bayesian method (Bond et al. 1989). And we only use the highest frequency channel data because the data of the other channels seem to be suffered the galactic emission (Gaier et al. 1992). We also take into account the subtraction of mean and gradient.

PIB models

Here the fully ionized universe is assumed. We check the models in wide range of parameters that are density parameter Ω, power law index n and non dimension Hubble constant h though the most preferable model from numerical simulations is $\Omega = 0.1 \sim 0.2$ and $n = 0.5$. We consider both open and flat universes without and with the cosmological constant, respectively.

The expected value of σ_8 and quadrupole anisotropies on the $\Omega - n$ plane are shown in Fig.3 (a) and Fig.3 (b) for $h = 0.5$ in the open universe. It is very impressive that σ_8 of the most desirable model is very close to unity. The expected

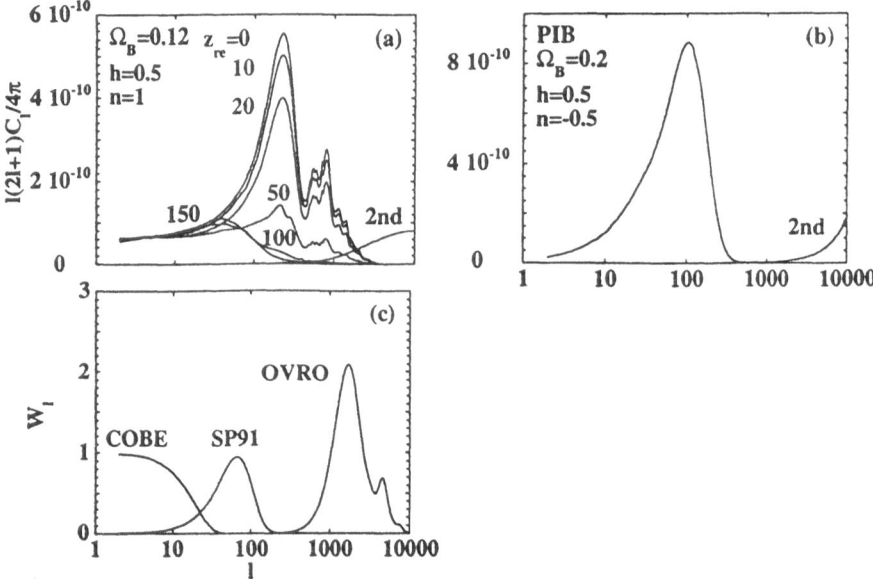

Fig. 2. Power spectrum of temperature anisotropies $\ell(2\ell + 1)C_\ell/4\pi$ as a function of ℓ for CDM with various reionization epochs (a) and PIB (b). Contribution of Vishniac effect is shown on large ℓ labeled as 2nd. We multiply the effect by factor 10 for CDM models. Window function W_ℓ for COBE, SP91 and OVRO are shown in (c).

values of quadrupole anisotropies are a little bit smaller than that of observed ones by COBE. But if we take into account of 'cosmic variance', these values are acceptable. In Fig.4, the constraints from the OVRO experiment, SP91 are shown for $h = 0.5$ in the open universe. About OVRO experiment, the right hand region of the 2.1×10^{-5} line is excluded. As for SP91, the right hand region of each line, labeled as $30\mu K$ and $25\mu K$ corresponds to the value of COBE normalization at $10°$ is excluded. Though the Vishniac effect is dominated, expected anisotropies on the arcminute scale are still a little bit lower than the observed ones. But the problem is the degree scale anisotropies. Because the new doppler peak just coincides with the South Pole experiment scale as shown in Fig.2(b), this experiment gives very stringent constrains on the models. As a result, not only these specific parameter models, but also almost of all models motivated on the PIB models are excluded.

CDM models

Here we only consider the *standard* models with $\Omega = 1$, $n = 1$ and initially adiabatic perturbations. We assume that the universe becomes suddenly ionized at the some epoch after the recombination. We consider the density parameter of baryon Ω_B, Hubble constant h and this epoch of reionization z_{re} as parameters. The expected values of σ_8 and the quadrupole anisotropy do not depend on the thermal history because CDM spectrum is changed very little due to the thermal history and the quadrupole anisotropy is far larger than a horizon scale of the

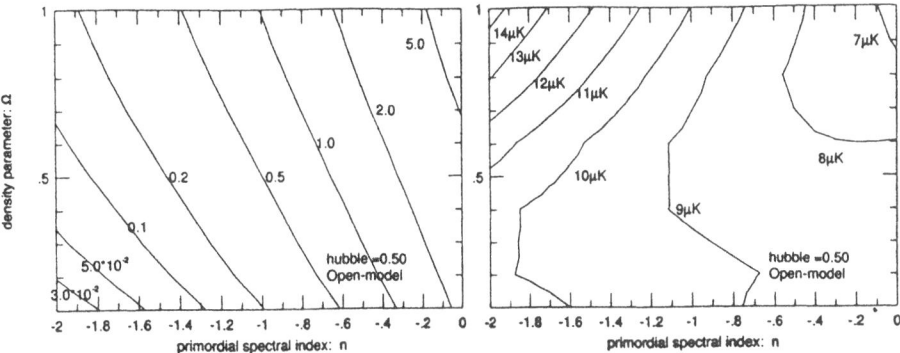

Fig. 3. The expected values of $\sigma 8$ (a) and quadrupole anisotropy (b) for PIB models by COBE normalization are shown in $\Omega - n$ plane.

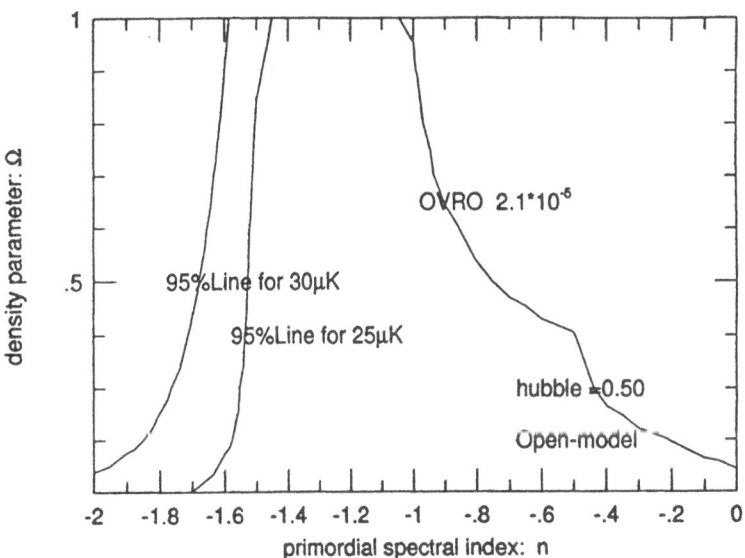

Fig. 4. The 95% credible-limit on PIB models from OVRO and SP91 experiment.

last scattering surface. And these values are consistent with observed ones. It is know that the expected arcminute scale anisotropies are still smaller than observed ones even in the non-reionized universe. If reionization is assumed, anisotropies on small scale are smeared out. So the observations on small scale do not give any constraints on models. As for CDM models, the Vishniac effect is not so important because the last scattering surface is father than that of PIB models. The Vishniac effect is suppressed by proportional to square of red-shift at the last scattering surface since it is a second oder effect. The constrains from SP91 on $\Omega_B - h$ plane are shown in Fig. 5. If we consider reionization at $z = 10$, the constraint is almost same as the models without reionization (Dodelson and Jubas). But it is very remarkable that reionization at $z = 20$ makes the constraint significantly looser and all models which are consistent with Big Bang Nucleosynthesis are survived though the universe is far from being optically thick at last scattering.

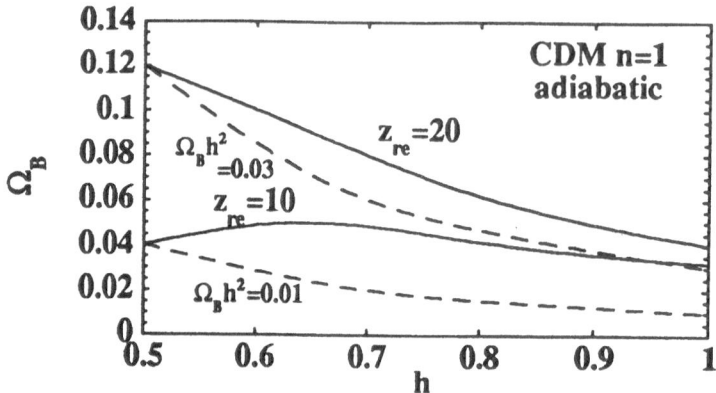

Fig. 5. Constraints on CDM models form SP91 in the $\Omega_B - h$ plane. The 95% credible-limit is shown. The epoch of reionization is chosen as either $z = 10$ or $z = 20$. Lines corresponding to $\Omega_B h^2 = 0.01$ and $\Omega_B h^2 = 0.03$ are also shown.

4 Conclusion

We have investigated effects of late time reionization on the CMB anisotropies. Such reionization smooths out the original temperature fluctuations on the scale smaller than a few degrees while it creates new fluctuations on large scale which corresponds to the horizon scale of the last scattering surface and on small scale by the Vishniac effect. We compared the expected anisotropies of PIB models and CDM models with several recent observations.

We found PIB models are almost dead if we assume the continuously ionized universe. We also have checked the constraint from the new 13 points experiment at South Pole by Schuster et. al (1993) and still get almost same answer. It seems that the end of PIB models can not be avoided. However late time reionization may save this model from death. The reason is if we consider the optically thin universe, the new peak on the spectrum of CMB anisotropies does not create and the Vishniac effect is small. The problem is the original doppler peak of CMB anisotropies. We will find the minimum fluctuations of PIB models and give the most stringent constraint in near future.

As for CDM models, we found that reionization only at $z = 20$ is required in oder to save the large Ω_B models which are consistent with Big Bang Nucleosynthesis though the universe is still transparent.

The CMB spectrum on the scale smaller than a few degree is very sensitive to the thermal history of the universe. So nowadays, it becomes more and more important for the experiments of the CMB anisotropies on the intermediate scale.

The author acknowledges financial support from a JSPS postdoctoral fellowship for research abroad. A part of this work about PIB models is collaborated with T. Chiba and Y. Suto and about CDM models is collaborated with J. Silk and N. Vittorio.

References

Bond, J.R., Efstathiou, G., Lubin, P.M., and Meinhold, P.R., *Phys. Rev. Letter.* **66** (1989) 2179.

Cen, R., Ostriker, J., and Peebles, P.J.E., (1993) *preprint.*

Dodelson, S. And Jubas, J.M., *Phys. Rev. Letter.* **70** (1993) 2224.

Gaier, T., et al., *Ap. J. (Letters)* **398**(1992) L1.

Gorski, K.M., Stompor, R., and Juszkiewicz, R., *Ap. J. (Letters)* **410** (1993) L1.

Ostriker, J.P., and Vishniac, E.T., *Ap. J. (Letters)* **306** (1987) L51.

Readhead, A.C.S. et al., it Ap. J. (Letters) **346** (1989) 566.

Schuster, J., et al., *Ap. J. (Letters)* **412** (1993) L47.

Silk, J., *Ap. J.* **151** (1968) 459.

Smoot, J., et al., *Ap. J. (Letters)* **396** (1992) L1.

Suginohara, T., and Suto, Y., *Ap. J.* **387** (1992) 431.

Sugiyama, N., and Gouda, N., *Prog. Theor. Phys.* **88** (1992) 803.

Vishniac, E.T., *Ap. J.* **322** (1987) 597.

Microwave Background Anisotropies: Future Plans

Paolo de Bernardis, Roberto Maoli*, Silvia Masi,
Bianca Melchiorri**, Francesco Melchiorri,
Monique Signore***, Danilo Tosti

* Laboratoire de Radiastronomie Millimetrique
Observatoire de Meudon, Unité CNRS-336
5, Place Jules Janssen 92195- MEUDON Cédex- FRANCE

** IFA-CNR Pzle L. Sturzo 31, Rome, Italy

*** Ecole Normale Superieure
24 Rue Lhomond F-75231 PARIS Cédex 0.5- FRANCE

Department of Physics Guglielmo Marconi
University of Rome "La Sapienza", Pzle Aldo Moro 2, ROME, Italy

Abstract

The role of CBR Anisotropies in Cosmology is briefly discussed. The various possible sources of noise and distortions are analyzed. In particular we consider the effects of the elastic Thomson scattering between Cosmic Background Radiation (CBR) and primordial molecules, like H_2, H_2^+, HeH^+, HD, HD^+, LiH, LiH^+ : primary CBR anisotropies are erased or attenuated for angular scales $\leq 10^0 \Leftrightarrow 50^0$ and frequencies $\nu \leq 50\ GHz$.

Future space programs are illustrated.

Subject headings: Cosmology- Cosmic Background Anisotropies- Post-Recombination Universe

1. Introduction

The potentiality of Cosmic Background Radiation (CBR) anisotropy studies lies on the hypothesis that the post-recombination Universe (i.e. $10 \leq Z \leq 1000$) was transparent to CBR photons so that measurements of anisotropies at various angular scales can be directly compared with the predictions of theories of galaxy formation. It is therefore required that the post-recombination Universe was not re-ionized, because an ionized medium would scatter CBR photons thereby changing their energy distribution and spatial direction. Strong evidence against an hot re-ionization (i.e. characterized by an electron temperature $\geq 10^6$ ^0K) is provided by recent COBRA (Gush et al. 1990) and COBE-FIRAS (Mather et al. 1990) results: their spectral measurements of CBR have set stringent upper limits to any spectral deviation from a pure Planckian curve as well as to the Comptonization parameter, excluding the existence of an hot ionized medium. If we neglect for a while the remaining possibility of a soft reionization ($T_e \leq 10^5$ ^0K) or a very late re-ionization (i.e. occurring at redshift $Z \leq 70$), we are led to the conclusion that CBR anisotropies, observed by COBE-DMR at angular scales from 90 to 10 degrees are directly testing primordial perturbations.

This being the situation let us briefly summarize the criteria employed by cosmologists to describe CBR Anisotropies:

The entire sky distribution of CBR may be described in terms of spherical harmonics:

$$T = \sum_{\ell=0}^{\ell=\infty} \sum_{m=-\ell}^{m=\ell} A_\ell^m Y_m^\ell(\Theta, \Phi) \qquad [1]$$

Where the first term of the sum $A_0^0 Y_0^0 = T_0 = 2.73K$ is just the CBR mean temperature, averaged over the sky. Let us assume that we explore a certain sky region: we are interested in the r.m.s. value of the temperature fluctuations. Therefore we have to compute the quantity $\langle (\frac{\Delta T}{T})^2 \rangle$. We may write:

$$\left\langle \left(\frac{\Delta T}{T}\right)^2 \right\rangle = \frac{1}{4\pi} \int \left(\frac{\Delta T}{T}\right)^2 (\Theta, \Phi) d\Omega =$$

$$= \frac{1}{4\pi} \sum_{\ell\ell'} \sum_{mm'} a_\ell^m a_{\ell'}^{'m'} \int Y_m^\ell(\Theta, \Phi) Y'_{m'}^{\ell'}(\Theta, \Phi) d\Omega \qquad [2]$$

The integral of spherical harmonics is zero unless $\ell = \ell', m - m'$, so that

$$\left\langle \left(\frac{\Delta T}{T}\right)^2 \right\rangle = \sum_\ell \sum_m \frac{\langle |a_\ell^m|^2 \rangle}{4\pi} \qquad [3]$$

In a uniform Universe we expect a rotational invariance of the coefficients, so that the sum over m simply gives $2\ell + 1$ and, finally, we get

$$\left\langle \left(\frac{\Delta T}{T}\right)^2 \right\rangle = \sum_\ell \frac{(2\ell + 1)C_\ell}{4\pi} \qquad [4]$$

Where the coefficients $C_\ell = \langle |a_\ell^m|^2 \rangle$ are the quantities we want to evaluate.

Theories of galaxy formation provide informations on the dependence of these coefficients on ℓ: the normalization, however, is mainly experimental. It is customary to plot the quantity $\zeta_\ell = \ell(2\ell + 1)C_\ell$ versus ℓ because in the most popular theory of Cold Dark Matter + Inflation ζ_ℓ is expected to be a constant for larger- than- horizon anisotropies. In Figure 1 we have plotted some of these "theoretical power spectra" just to give an idea of the variety of the possible theories.

In this figure we have also sketched the power spectra of the most significant sources of disturbances; namely, atmospheric fluctuations, galactic dust emission, radio-galaxies counts.

Let us investigate the effects of the instrumental limitations. In the practical cases, one has to introduce an instrumental function $F(\ell, \Theta, \sigma)$ which depends on the beam size σ (for a Gaussian shape of the angular response, σ is just the variance), on the beam modulation amplitude (two or three sky regions are observed, separated by an angle Θ) and on the type of modulation (two beams, three beams, sinusoidal, square-wave, etc) and demodulation. In the two simple cases of square-wave, two and three beams modulation the instrumental function is of the type

$$F(\ell)_{two-beams} = 2[1 - P_\ell(\cos \Theta)] \exp\left[-\left(\ell + \frac{1}{2}\right)^2 \sigma^2\right] \qquad [5]$$

$$F(\ell)_{three-beams} = \left[\frac{3}{2} - 2P_\ell(\cos \Theta) + \frac{1}{2}P_\ell(\cos 2\Theta)\right] \exp\left[-\left(\ell + \frac{1}{2}\right)^2 \sigma^2\right] \quad [6]$$

Where $P_\ell(\cos \Theta)$ are the Legendre polynomials: for large ℓ they approximate the Bessel function $J_0((\ell + 1/2)\Theta)$. In Figure 1 we have illustrated how the Transfer Function changes for different experiments: they would probe different regions of the power spectrum and, as an obvious consequence, also different physical mechanisms responsible for the anisotropies.

Unfortunately, as more data become available from new intermediate-scale experiments, it appears more and more evident that none of the various available theories is fitting the data satisfactorily.

For instance, Tenerife (Lasenby et al., 1993) and South Pole Experiment (Schuster et al. 1993) seem to indicate upper limits well below the values expected by a simple extrapolation from COBE-DMR data on the basis of Cold Dark Matter models (Vittorio, 1993). This situation has led some authors to revitalize the hypothesis of a moderate secondary reionization around $Z \simeq 100 - 50$ in order to appropriately reduce intermediate-scale anisotropies to values compatible with CDM theory, while avoiding spectral distortions within the limits posed by COBE-FIRAS results (Silk et al., 1993).

Even more disturbing is the fact that bolometric observations in the millimetric region at similar angular scales, like one of the ULISSE experiments (Melchiorri et al. 1981), the MAX experiment (Devlin et al. 1992) and ARGO experiment (de Bernardis et al. 1993) provide r.m.s. values more than 2σ higher than the radio results. The disagreement between radio and infrared measurements could well be fortuitous, being the observations carried out on

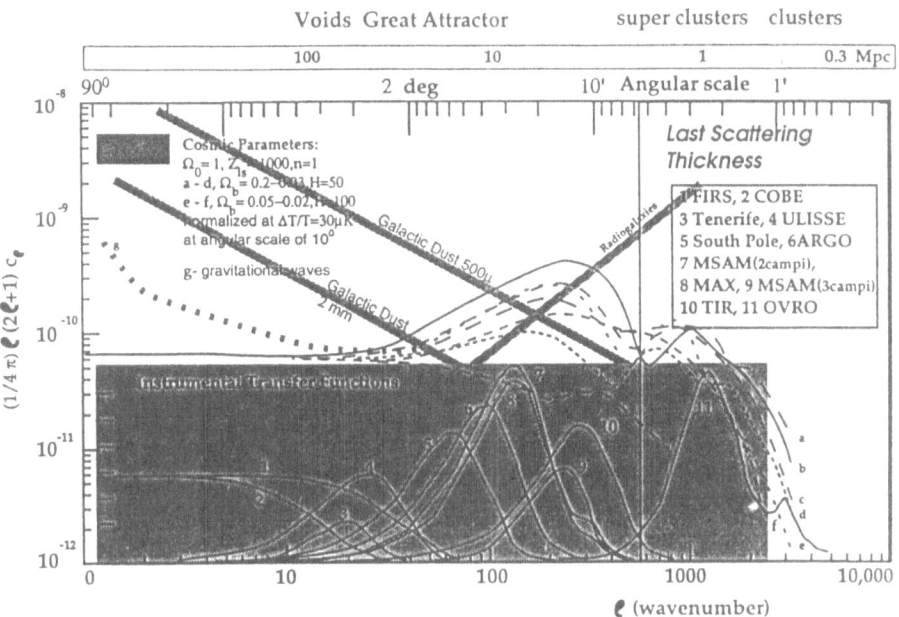

Fig. 1: Power spectrum of CBR Anisotropies as expected in the case of CDM+inflation with and without reionization. Galactic dust spectrum is shown as deduced from ULISSE, ARGO and MAX experiments

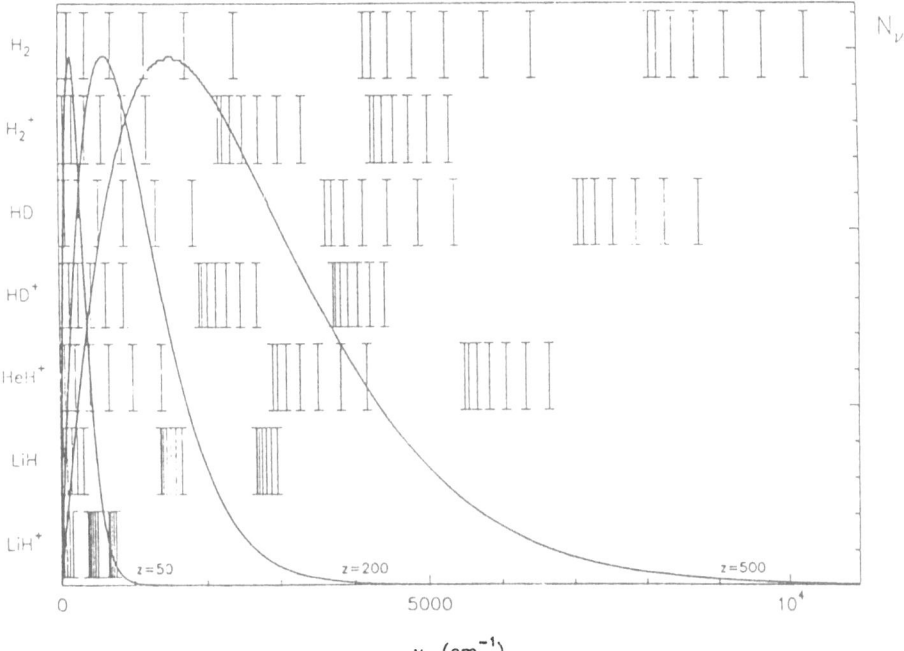

Fig. 2: Energy levels of the considered molecules. The black-body photon density normalized to its maximum value is plotted for three different redshift.

different sky regions : in such a case, however, one is forced to consider a non-Gaussian spatial distribution of anisotropies.

Therefore, it appears worthwhile to re-analyze the general question of the cosmic transparence to CBR photons in order to explore the possibility that CBR anisotropies could be affected by some not yet considered mechanism, even in the case of a neutral Universe. In view of the problems risen by the observations, such an unknown mechanism should present the following peculiarities:

a- it should not affect CBR spectrum.

b- It should be able to erase, at least partially, CBR anisotropies at small angular scales and, possibly, in a frequency-dependent way able to explain the colour excess in the millimetric region.

c- secondary anisotropies arising from the process should pass the severe upper limits already posed by several radio and infrared observations at small angular scales.

The goal of the present paper is to show that points a) to c) may be fulfilled by the process of elastic Thomson scattering between CBR photons and primordial molecules : more specifically, we intend to study the influence of primordial molecules on CBR anisotropies in the redshift range $Z_f \Leftrightarrow Z_d$, where Z_f is the *formation redshift*, i.e. when the Universe becomes cold enough to stop the photodissociation of newly formed molecules ($Z_f \simeq 100 - 400$ depending on the molecular species); Z_d is the dissociation redshift, i.e. when the high energy photons of the first formed objects start again to photodissociate molecules ($Z_d \simeq 5 - 70$ depending on the evolutionary models of structures).

In the above redshift range a certain amount of primordial molecules is expected to be present. The possible mechanisms of interaction between these molecules and CBR photons are thermal emission and absorption and elastic Thomson scattering. Since molecular abundances are modest, collisional excitation and de-excitation are negligible: therefore, thermal emission and absorption are also negligible. Puy et al. (1992) have estimated that emission from molecules could hardly reach 10^{-6} of CBR brightness.

The remaining effect is the Thomson scattering between photons and molecules. This is an elastic scattering during which a photon is first absorbed and then re-emitted at the same frequency, but not in the same direction. Obviously, the effect does not alter CBR spectrum,while it could in principle smear out the primordial spatial distribution of CBR: For a given molecular line ν_{ij} a CBR photon arriving from the Last Scattering Surface can be resonantly scattered if $\nu_{LS}\frac{(1+Z)}{(1+Z_D)} = \nu_{ij}$; taking into account the finite natural linewidth $\Delta\nu_{ij}$ (essentially due to the Doppler broadening), this condition will be fulfilled for a redshift range $\frac{\Delta Z}{Z} = \frac{\Delta\nu_{ij}}{\nu_{ij}} \simeq 10^{-5}$.

In the following paragraphs we summarize our knowledge about molecular formation in the hostile pregalactic environment, compute the optical path, analyse the effect on CBR anisotropies and, finally, briefly discuss the possible secondary anisotropies .

2.- Molecular Formation in Early Universe and Optical Depth Computation

2.1 Primordial abundance of molecules :

Molecular formation in the early Universe starts when photons have not enough energy to dissociate them. The final abundance for each specie depends on the initial atomic abundance, and on the associative and dissociative rates of the various reactions. A complete lists of possible reactions can be found in Lepp and Shull (1987), Kirby and Dalgarno (1978), Dalgarno and Lepp (1987), Puy et al. (1992).

Of particular relevance is the case of molecules containing Lithium: LiH and LiH^+. The abundances of these molecules are rather uncertain: for LiH the data available in literature range from $1 - 10\%$ of the primordial ^7Li abundance (Dubrovich,1977; Lepp and Shull, 1984) to 60% (Puy et al. 1992). More recently, Lipovka (1993) has shown that almost all the available Li is trapped in LiH, if one correctly takes into account the temperature dependence of the photoassociation rate. For LiH^+ the situation is even more uncertain: since ionized Li recombines later than hydrogen, Dalgarno and Lepp (1987) concluded that LiH^+ could even be more abundant than LiH and Palla and Fink (1992) estimated an abundance of about 50% of Li. The rates for the involved chemical processes are almost unknown, however.

In any case, the values theoretically expected for LiH, range from 6×10^{-10} predicted by standard nucleosynthesis (Yang et al. 1984; Kawano et al. 1988; Arnould and Forestini, 1989), up to a maximum of 10^{-7} in the case of inhomogeneous nucleosynthesis (Reeves et al. 1988, 1990; Kurki-Suonio et al. 1988, 1989). We want to point out that primordial Li abundance is related to the baryon-photon ratio: the uncertainties in this number allow a range from 10^{-10} to 10^{-9} even in the framework of the standard Nucleosynthesis. Measurements toward Population II stars tend to a low value like $1 - 2 \times 10^{-10}$ (Spite and Spite , 1982; Rebolo et al. 1988), but these results have been criticized by various authors , being possible that a substantial fraction of Li was destroyed through some form of internal mixing (Vauclair 1987, 1988). The most recent measurements of Li abundance is that of Lemoine et al (1992) giving an abundance of 3.4×10^{-9} toward $\rho - Ophiuchi$. In any case, in Table I we have indicated the adopted ranges of abundances.

Table I

Molecule	abundance	Z_f
H_2	$10^{-6} \div 10^{-5}$	355
H_2^+	$10^{-13} \div 10^{-12}$	210
LiH	$.1 \div .9 \cdot [Li]$	400
LiH^+	$\leq .1 \cdot [Li]$	> 400
HeH^+	$\sim 10^{-14}$	148
HD	10^{-11}	357
HD^+	10^{-10}	212

2.2- Optical depth computation

The molecules we considered are H_2, H_2^+, LiH, LiH^+, HeH^+, HD, HD^+: their energy levels are shown in Figure 2 .

The resonance cross section of Thomson scattering has to be normalized on the line width and the optical depth τ is

$$\tau = \int \sigma n_{mol} d\ell \qquad [7]$$

where

$$\sigma = \frac{\lambda^3 A_i j}{8\pi c} \frac{\nu}{\Delta\nu_D}$$

$$n_{mol} = \Omega \omega_b \rho_c \alpha_m n_{vj} (1+Z)^3$$

$$d\ell = \frac{cdZ}{H_0(1+Z)^2\sqrt{1+\Omega Z}}$$

where the notations mean:

σ = cross section;

$\Delta\nu_D$ = Doppler broadening;

A_{ij} = Einstein coefficients;

Ω =total (i.e. baryonic and non-baryonic) density relative to ρ_c ;

ω_b = baryonic fraction of total density

ρ_c = critical density; α_m = molecular abundance to H;

n_{vj} = population of level with quantum numbers v, j

The abundance of CBR photons guarantees that the population of the levels follows the Boltzmann distribution, at least for $Z \geq 70$: for smaller values of Z one has to consider the collisional processes, but the correction is small in our case.

Due to the expansion of the Universe a photon will change its frequency and explore the entire linewidth of a given molecular resonance within a redshift interval $\Delta Z = (1+Z) \times \frac{\Delta\nu}{\nu}$, where $\Delta\nu$ is the linewidth, essentially determined by the Doppler broadening. The optical depth will depend on the molecule abundance and on the number of molecules in each quantum level considered as starting level in the process of elastic scattering. The optical depths for rotational and vibrational transitions are shown in Figure 3 a,b,c,d.

Even if H_2 is the most abundant, its interaction with photons is small, due to the lack of dipole moment: moreover the quadrupole coupling is not sufficient to provide interesting scattering. Similarly H_2^+ (lacking of dipole moment too) and HD (having $\mu_D \simeq 10^{-4}$ esu) give a negligible contribution and are not plotted in Figure 4: the only important contributions arise from LiH and LiH^+.

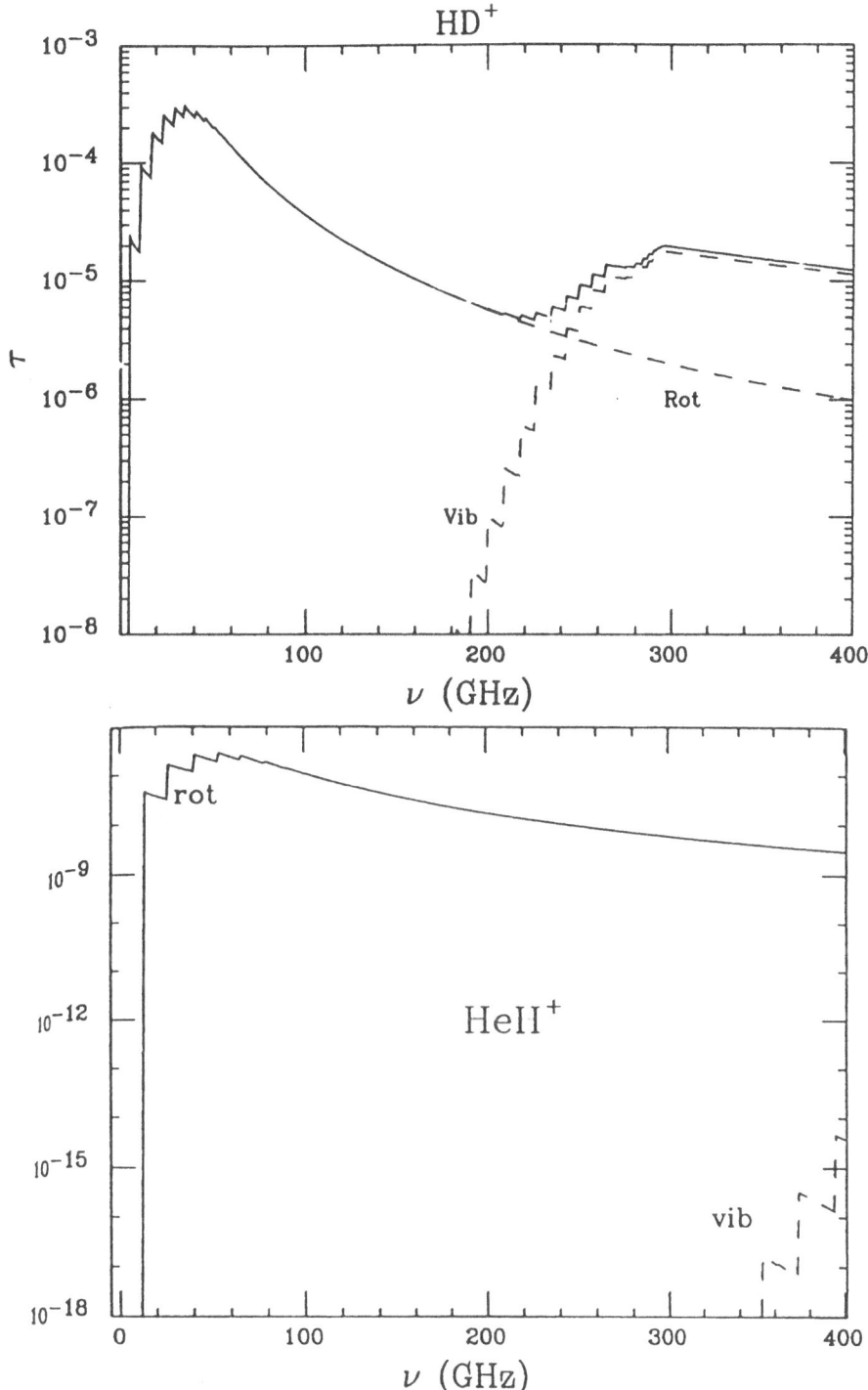

Fig. 3a, b: Optical depth due to each molecule as a function of redshift; we plotted separately the rotational and vibrational transitions contribution (dashed lines).

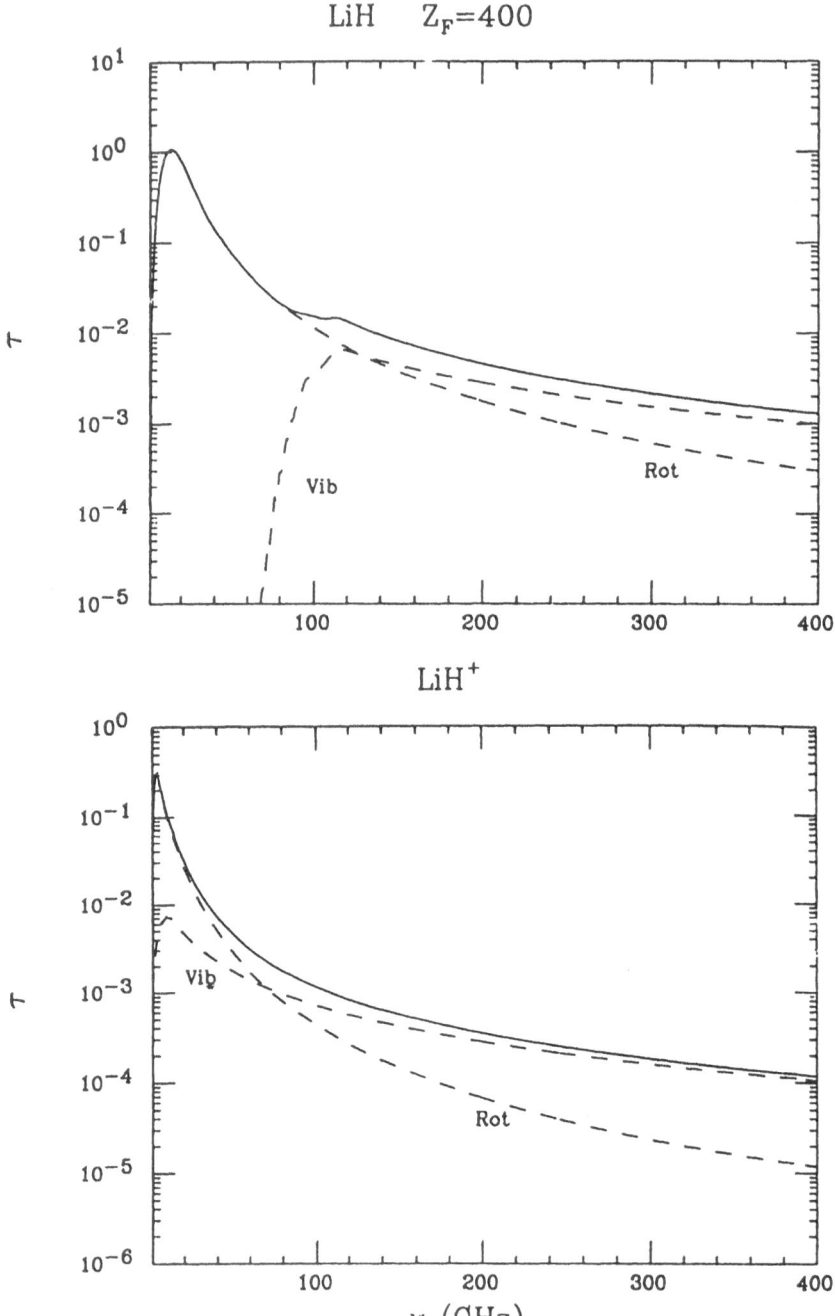

Fig. 3c, d,: Optical depth due to each molecule as a function of redshift; we plotted separately the rotational and vibrational transitions contribution (dashed lines). For LiH we considered $Z_f = 400$ and $[LiH] = 3 \cdot 10^{-9}$. For LiH^+ we considered $[LiH^+] = 3 \cdot 10^{-9}$.

The total optical depth is shown in Figure 4 a,b,c,d. We have considered the most significant cases, namely LiH formation at redshift 200 and 400 with a redshift of photodissociation respectively 5 and 70. For each molecular scenario, the total optical depth is computed in the case of a large Li abundance (like in the case of inhomogeneous Nucleosynthesis), Population II abundance and standard Nucleosynthesis. For comparison, the frequencies of COBE-DMR radiometers are indicated .

3- Damping of Primary CBR Anisotropies; Rising of Secondary Anisotropies

The first important consequence of Figure 5 is that primary CBR anisotropies may be affected by resonant scattering if $\Theta \leq \Theta_H$ and $\nu \leq 80 GHz$, where Θ_H is the angular diameter of the Horizon at the effective redshift of Thomson scattering. In order to get an estimate of Θ_H we recall that the angle subtended by the horizon at a redshift Z is given by

$$\Theta_{horizon} = 2 \arcsin \left[\frac{\Omega_0 (\Omega_0 Z + 1)^{1/2}}{\Omega_0 Z + (\Omega_0 - 2) \left[(\Omega_0 Z + 1)^{1/2} - 1 \right]} \right] \qquad [8]$$

we averaged all the horizon diameters for the lines contributing at different redshift to the same frequency, i.e.

$$\Theta_H(\nu) = \frac{\sum \Theta_H^i \tau_i}{\tau_i} \qquad [9]$$

where τ_i is the optical depth of the single line. The dependence of Θ_H on the frequency is shown in Figure 5 a,b in the two cases of dissociation redshift 5 and 70. For a given power spectrum of primary CBR anisotropies, the effect of the Thomson scattering is that of affecting all the angular scales smaller than $\Theta_H(\nu)$ by an amount of the order of $e^{-\tau}$. Therefore we anticipate a decrease in CBR anisotropies which is wavelength dependent, as shown in Figure 6 a, b. For sake of simplicity we have estimated the only two cases only of inhomogeneous Nucleosynthesis and standard Nucleosynthesis.

As primary CBR anisotropies are erased, new, secondary anisotropies are expected. We may consider three classes of secondary anisotropies; they all arise from the primordial clouds evolving from the perturbations present at the last scattering surface.

To get an idea of the peculiarities of these anisotropies let us first consider the spectrum of a single cloud located at different redshifts but with mass corresponding to a galaxy cluster today (this fixes the peculiar velocity at about 600 Km/sec now). In Figure 7 we have plotted the corresponding spectrum . In order to isolate a single cloud along the line of sight, the observer should use a telescope with an adequate spatial and spectral resolution. For a cloud with dimensions comparable with a proto-galaxy, the spatial resolution would range

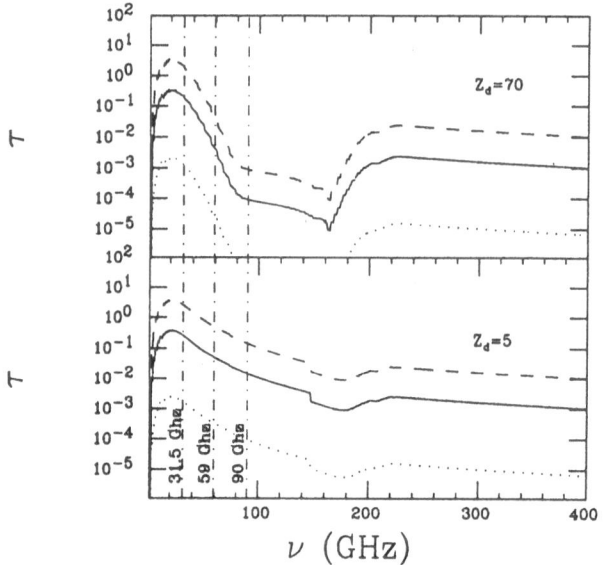

Fig. 4a, b: Total optical depth due to all molecules. We considered two different cases for the LiH formation redshifts and three cases for the initial lithium abundance: standard nucleosinthesys (dotted lines), popolation II stars (continous lines) and inhomogeneous nucleosinthesys lithium abundance (dashed lines). It is fixed $Z_f = 200$.

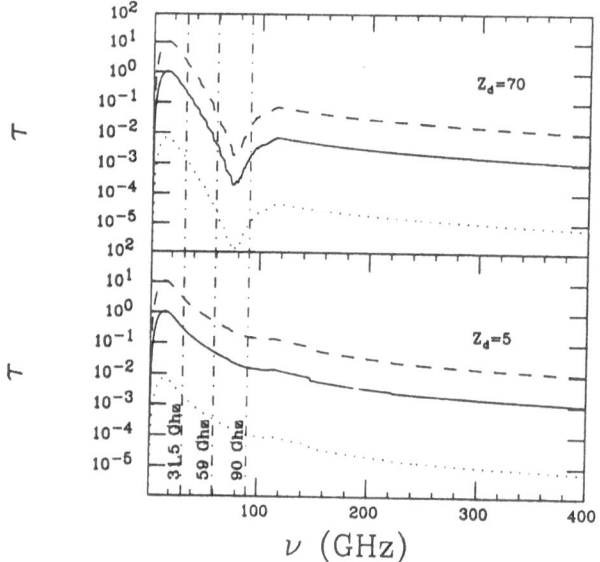

Fig. 4c, d: As in figures 4a and 4b for $Z_f = 400$.

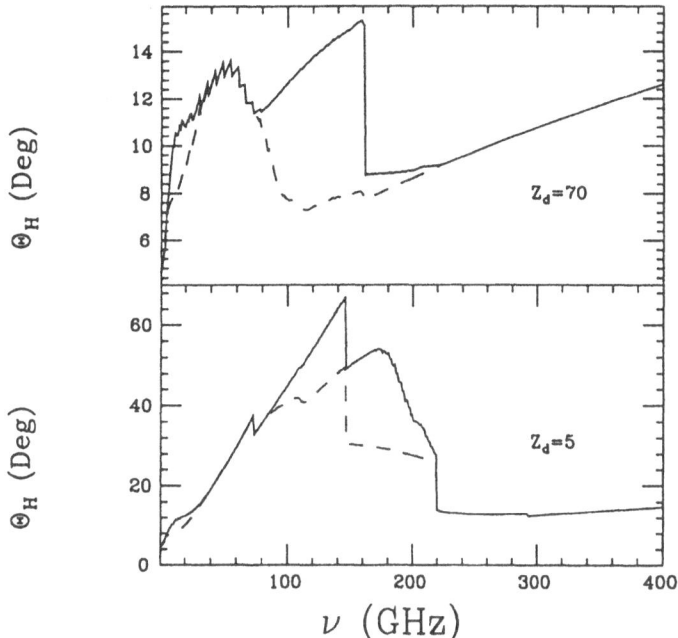

Fig. 5a, b: Horizon angular dimension as a function of redshift. We considered two different re-ionization redshifts, as indicated. Dashed lines are calculated for $Z_f = 400$ and continous lines for $Z_f = 200$.

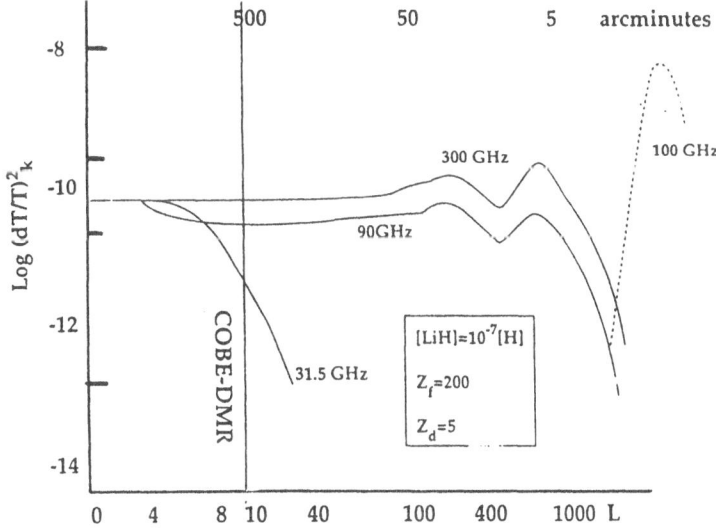

Fig. 6a: Anisotropies power spectrum per logaritmic interval. We plotted them at different frequencies considering the cases of the greatest optical depth we got, as indicated. The secondary anisotropies raising from non-isotropic scattering are also shown in the figure (dashed line). The continuos vertical line refer to COBE-DMR experiment angular scale.

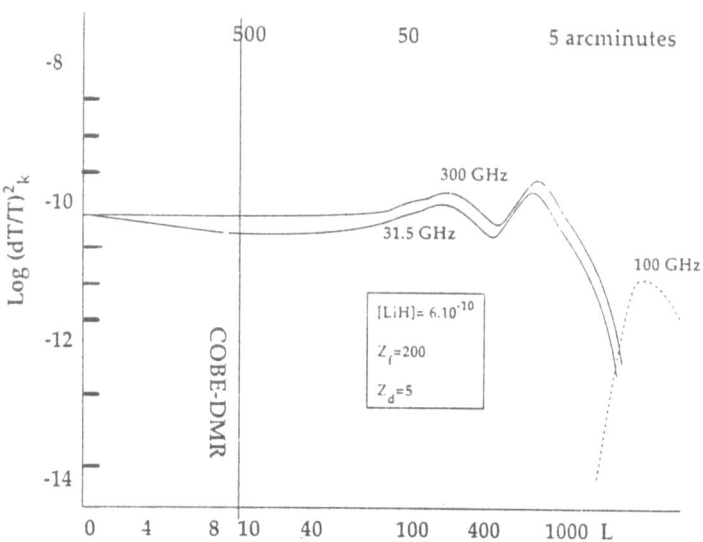

Fig. 6b: As in Figure 6a for the lowest obtained optical depth.

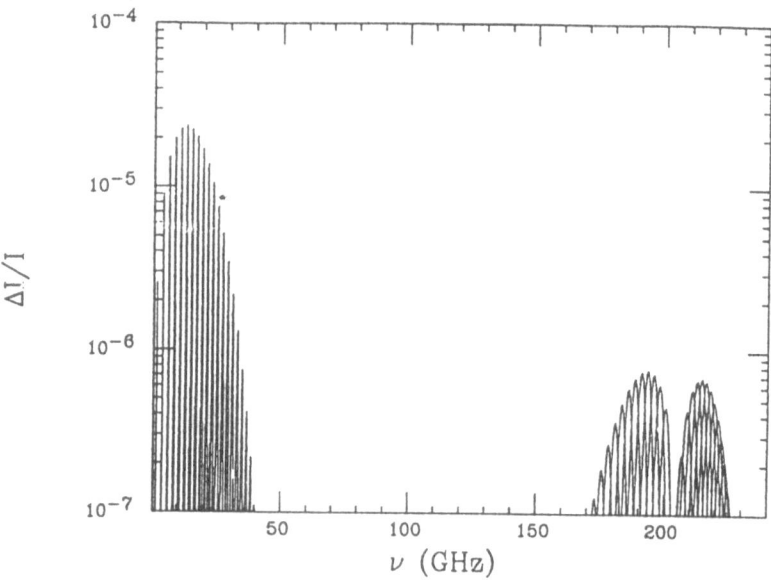

Fig. 7: Theoretical spectrum for a typical isolated cloud of galaxies cluster size at $z=200$, moving at peculiar velocity $v_p = \frac{v_{p0}}{\sqrt{1+z}}$. The resulting spectrum is essentially due to LiH, with a small contribution due to LiH^+ (bold lines).

around 10-50 arcseconds for a redshift $Z \simeq 100$, while the spectral resolution is given by $\Delta \nu \leq \nu \frac{\Delta Z}{Z}$ in the case (a,b) and about 10^{-5} in case (c). Even in these conditions one should take into account that other clouds along the line of sight at appropriate redshifts would radiate inside the selected bandwidth and, also, would scatter the radiation coming from the selected cloud. In Figure 8 a, b, c, d the expected anisotropy is plotted versus the frequency of observation in the same cases as for the optical depth.

We should note that this is a rough estimate of the real anisotropy, to which clouds of different size contribute. However it is adequate to provide an idea of the level of the expected signals for clouds in linear regime. The angular size of these anisotropies would be that of protogalactic objects, i.e. 10"-100". The main conclusion is that secondary anisotropies have a very peculiar spectrum: In the range $20 \leq \nu \leq 250 \, GHz$ they may acquire detectable amplitudes. Moreover, in the region $\nu \geq 60 GHz$,the Universe is rather transparent and $\tau << 1$, and one should be able to observe the effect of the single cloud. As pointed out before this would require a spectral resolution $\frac{\Delta \nu}{\nu} \simeq \frac{\Delta Z}{Z} \simeq 10^{-3}$ for the typical dimensions of clouds corresponding to galaxies today($\simeq 10$ arcseconds). Moreover, the spectral distribution of the roto-vibrational lines would allow to check the nature of the observed signals, being the roto-vibrational signature unique in this range of frequencies.

A rather different situation is expected in the case of clouds at the end of the linear regime: as $\frac{\Delta \rho}{\rho}$ approaches the unity, the cloud collapse compensates the expansion of the Universe. The collapsing velocity is close to $\frac{c \Delta Z}{(1+Z)}$ and the expected signals are remarkably large, even if concentrated into narrow lines. (see Appendix A).

4. Comparison With Observations

Our model predicts a partial or total erase of anisotropies in the radio region for angular scales smaller than the Horizon at the redshift for which the optical depth is unity. This effect is similar to the case of a partial secondary ionization but differs in being strongly frequency dependent. If primordial Li abundance is large (i.e. $\geq 10^{-9}$) we predict a substantial difference between the amplitude of COBE-DMR 31 GHz Channel and 90 GHz Channel. Therefore, one may use COBE-DMR data to set upper bounds to Li abundance. Unfortunately, the amplitude of 31 GHz channel is the most noisy in COBE data and the observed anisotropy could even be consistent with zero at this frequency. (Smoot et al. 1992). Much more sensitive to any possible erasure are the various 30 GHz experiments carried out with HEMT's at South Pole (Lubin et al., 1993) and at Saskatoon (Wilkinson et al. 1993) at angular scales of 0.5-1.5 degrees. The results of these experiments, although very preliminary, seem to indicate an anisotropy at a level of 1.5×10^{-5}, significantly (2σ) smaller than the value quoted by MAX experiment (Richards et al., 1992) at 200 GHz. If this discrepancy will be confirmed on the same sky region, then the effect of primordial molecules would be proved. At the present level of knowledge, however, the difference could well be fortuitous:

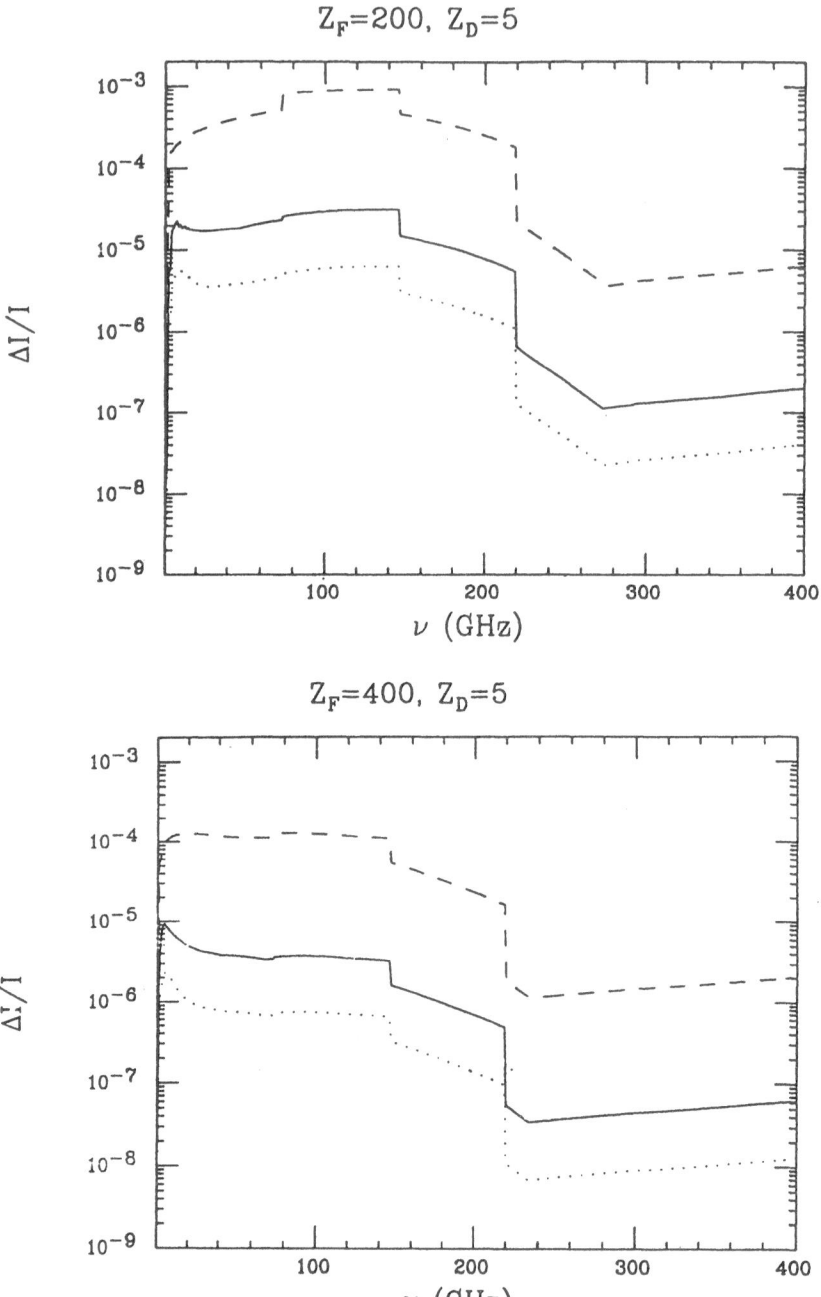

Fig. 8a, b: For a given frequency we considered the molecular lines giving the highest signal that we plotted after considering the action of other lines. The maximum signal is given by LiH rotational transitions from 10 GHz til 100-200 GHz (this last value depending on the re-ionization redshift Z_d.

We may tentatively summarize the observational situation as following:

1- Tenerife Experiment: Beam-size= 5.6 ^{0}, 15 GHz radiometer $\Rightarrow$ 60 ± 12μK; 37 GHz radiometer $\Rightarrow$ 86 ± 20μK. For a redshift of formation of $Z_F = 400$ we get $[LiH] \leq 2 \times 10^{-9}$.

2- SP and MAX Experiments: Beam size $\simeq 1^0$: 30 GHz radiometer $\Rightarrow \leq 50\mu K$; 200 GHz bolometer $\Rightarrow 100 - 150\mu K$. For a redshift of formation $Z_F = 400$ we get $[LiH] \leq 2 \times 10^{-9}$.

3- ULISSE and COBE: Beam size $\simeq 6^0$, 300 GHZ radiometer$\Rightarrow$ 110 ± 50μK; 31 GHz receiver, 10^0 beam size, $\Rightarrow \leq 50\mu K$. For a redshift of formation $Z_F = 400$ we get $[LiH] \leq 4 \times 10^{-9}$.

Secondary anisotropies may be tested at small angular scales, like in the radio observations of Fomalont and Partridge (1992) at 8 GHz and 50", although their spectral resolution of 50 MHz is a bit too low. Their upper limits of 2×10^{-5} impose a LiH upper limit of 3×10^{-9}. Observations at 200 GHz have been carried out by de Bernardis et al. (1992) and recently improved in sensitivity, by means of IRAM radiotelescope at 10 arcseconds resolution. The sensitivity of the system (about 1mK) is close to the values predicted by inhomogeneous Nucleosynthesis.

5- Conclusions

Primordial molecules may play a significant role in altering the amplitude and power spectrum of CBR anisotropies: the effect depends essentially on the Li abundance.

Conversely, accurate measurements of anisotropies at large and small angular scales will provide precious information on the abundance of Li: the upper limits we may derive from the available data are already excluding LiH values much greater than 10^{-9}: a further increase in accuracy of observations would possibly rule out inhomogeneous Nucleosynthesis.

Secondary anisotropies arising from anisotropic scattering, if detected, would represent the only way to study the spatial distribution of perturbations after the decoupling and well before the formation of non-linear structures.

This being the situation, it is hard to decide if signals observed by COBE and other groups are CBR anisotropies or a mixture of CBR and spurious signals : in the first case, their knowledge will provide a picture of the last scattering surface for angular scales greater than 10 arcminutes: at smaller angular scales the optical thickness of LSS is large enough to average and erase the anisotropies along the line of sight.

Therefore, what the study of CBR is providing at best is a bidimensional map of perturbations with comoving dimensions larger than those of super-clusters today. Referring to Figure 1 it follows that a suppression of angular scales smaller than 5 arcminutes (superclusters) would substantially leave un-changed the amplitude of the observed CBR anisotropies: a Universe without galaxies would produce the same anisotropies of the observed one! Viceversa, if we suppress all the large-scale fluctuations, so that CBR anisotropies become undetectable, the apparent distribution of galaxies would substantially remain unchanged; i.e. the observed nearby Universe could exist even with CBR anisotropies equal to zero. Still cosmologists pretend to reconstruct the process of galaxy formation from the analysis of this map. First of all it implies that the power spectrum of perturbations has a single slope from superclusters to galaxies, so that the study of the large perturbations on LSS may allow to predict the behaviour of the smaller one. The second point relies on the application of methods of analysis which have already failed in the study of galaxy distribution: the two point correlation function, when employed in galaxy statistics, was unable to emphasize important effects, like huge voids or matter concentrations. Why we have to believe that in the case of CBR this algorithm will provide a more efficient method?

The third point is that molecules, dust and gas may affect in several ways CBR photons along their travel from LSS to us. Primary anisotropies may be partially or totally erased in a wavelength dependent way as well as in a different amount depending on the angular scale. Secondary anisotropies may appear at small and intermediate angular scales.

This being the situation it is hard to believe that rapid progresses in Cosmology can be obtained, unless a wide systematic program of research is organized. In this program future measurements of CBR anisotropies may or may not play a relevant role, depending on the transparency of the Universe. Several satellite and balloon programs have been proposed to measure CBR anisotropies at 0.5-5 degrees : It is unfortunate that this interesting research program is apparently not complemented with what should be the natural evolution of observational Cosmology, i.e. the study of the Universe at $1000 > Z > 5$.

But let us analyse the situation in deeper detail. In our opinion the first and most important problem to be solved is that of the physical mechanism responsible for galaxy formation. Roughly speaking, two alternatives are still passing the observational tests: the linear theory of gravitational instability and the highly non-linear theory of matter condensation in the potential wells of topological defects, like Cosmic Strings. The last hypothesis is considered by some authors as a bit too odd to be taken seriously: one should remember, however, that nobody has observed the Hot or Cold non-baryonic matter required by the first hypothesis!

Cosmic Strings predict, however, a significant not-Gaussian distribution at small angular scales. Dedicated telescopes and special topological analysis are needed to solve the problem. It is unfortunate that all the proposed space experiments have not enough angular resolution to provide useful informations. A three meters telescope operating in the millimetric region would be adequate. The only available telescope is that of TIR Project, which however is at present in stand by due to an unfortunate decision of the Italian Space Agency. FIRST, the European millimetric space telescope could provide important informations, if bolometers will be used.

Once this problem will be solved, if the Theory of Gravitational Instability would prevail, the next important point is that of deciding the role of *Tensor Fluctuations*, i.e. of primordial gravitational waves. The amount of these perturbations is linked to the specific nature of the Inflation. As shown in Figure 1 this problem may be solved by a comparison between the amplitude of anisotropies at large (Quadrupole) and intermediate (2-5 degrees) angular scales, in order to check if the long-wave power spectrum is flat or not. COBE has provided a very limited amount of informations on this respect, just because the error bars are too large. The only planned space experiment in this angular range is RELIC 2. Some other proposed experiments, like SAMBA, would explore the intermediate-scale range, but a decision about GW contribution would require a new more precise measurement of the large-scale anisotropy.

The next problem to be solved is that of a possible re-ionization: the lack of a significant increase of anisotropies for angular scales smaller than the horizon at decoupling ($\leq$ 2-5 degree) would be a definite test of this fact. A wide range of experiments (balloon- satellite) are in the position to answer this question. One should note, however, that the presence of a significant reionization would render impossible CBR anisotropies observations at small angular scales. Among the various proposed space experiments only SAMBA has taken seriously into consideration this possibility, by including in the program a search for the S-Z effect on the ionized protoclouds, characterized by the typical spectral signature of SZ effect. Also, ground based observations of secondary anisotropies, like the VLBI searches could help in understanding the structure of the primordial Universe.

If reionization would turn out to be negligible, one has still to worry for molecular scattering: a comparison of CBR anisotropies at the same small angular scale but in the radio and millimetric regions would fix this question: it could also help in measuring the primordial abundance of Li. Search for secondary anisotropies by means of ground based telescopes and by FIRST could open the new field of cosmological spectroscopy.

In conclusion, the most promising results are expected to come from a combination of sub-millimetric and radio observations of CBR anisotropies at small angular scales: confirmations of COBE results are needed, however at much better level of sensitivity.

The suggestions and comments of P. Encrenaz have to be gratefully mentioned. Many of the ideas exposed in this paper have been originated by discussions with V. Dubrovich.

The research program was supported by MURST Grant 40% and CEE Twinning Program.

REFERENCES

Arnould, M. & Forestini, M., 1989 in *Proceedings of La Rabida School on Nuclear Astrophysics*, Springer Verlag, Berlin.

Dalgarno, A. & Lepp, S. 1985, in Astrochemistry Eds. Tarafdar, S. P., & Varshni, M.P., Reidel, 109

de Bernardis, P., Dubrovich, V., Encrenaz, P., Maoli, R., Masi, S., Mastrantonio, G., Melchiorri, B., Melchiorri, F., Signore, M. & Tanzilli, P.E.: 1993, Astron. Astrophys., **269**, 1.

de Bernardis, P., Aquilini, E., Boscaleri, A., De Petris, M., D'Andreta, G., Gervasi, M., Kreysa, E., Martinis, L., Masi, S., Palumbo, P., Scaramuzzi, F., "Degree-scale Observations of Cosmic Microwave Background Anisotropies", submitted to Ap. J.

Devlin, M., Alsop, D., Clapp, A., Cottingham, D., Fisher, M., Gundersen, J., Holmes, W., Lange, A., Lubin, P., Meinhold, P., Richards & P., Smoot, G. 1992, Proceed. Nat. Acad. Science, USA

Dubrovich, V.K. 1975, Pis'ma Astron. Zh., **1**, 10

Dubrovich, V.K. 1981, Izvestiya Spetsial'noi Observatorii , **13**, 40

Dubrovich, V.K. 1992 *Preprint* "Bluring of Primary Fluctuations by Molecules Curtain"

Fomalont E.B., Partridge R.B., Lowenthal J.D. & Windhorst R.A. 1993 Ap. J. **404**

H.P. Gush, M. Halpern & E.H. Wisshnow 1990 Phys. Rev. Lett. , **65**, 537

Kawano, L., Schramm, D., Steigman, G. 1988, ApJ., **327**, 750.

Kirby, K. & Dalgarno, A. 1978, Ap.J. **224**, 444.

Kurki-Suonio, H., Matzner, R.A., Centrella, J.M., Rothman, T. & Wilson, J.R. 1988, Phys. Rev. D., **38**, 1091.

Kurki-Suonio, H., & Matzner, R.A. 1989, Phys. Rev. D., **39**, 1046.

Lasenby, A.N., Davies, R.D., Rebolo, R. & Beckman, J.E. 1993, Proceedings of Santander Workshop on CBR.

Lemoine, M., Ferlet, R., Vidal-Majar, A., Emerich, C. & Bertin, P. 1992, submitted to Astr. Ap.

Lepp, S., & Shull, M. 1984, Ap.J. **280** 465

A. Lipovka 1993 *Private Communication*

Mather, J.C., et al. 1990 Proceedings University of Maryland Workshop "After the First Three Minutes"

Melchiorri, F., Olivo, B., Ceccarelli, C. & Pietranera, L. 1981, Ap.J. Lett. **227**, L129.

Palla, F. & Fink, E. 1992, I Convegno Nazionale di Cosmologia, University of Torvergata, Proceedings

Puy, D., Alecian, G., Le Bourlot, J., Le Orat, J. & Pineau, G., 1992 Astron. Astrophys., **267**, 337

Reeves, H., Delbourgo-Salvador, P., Audouze, J. & Salati, P. 1988, J. Phys., **9**, 179.

Reeves, H., Richer, J., Sato, K., & Terasawa, N. 1990, Ap.J. **355**, 18.

Rebolo, R., Molaro, P. &, Beckman, J.E. 1988, Astr. Ap., **192**, 192.

Schuster, J., Todd, G., Gundersen, J., Meinhold, P., Koch, T., Seiffert, M., Wuensche, C.A., & Lubin, P. 1993, Preprint University of California at Santa Barbara

Silk, J., Scott, D. & Wayne, H. 1993, submitted to Phys. Rev. D

Spite, F. & Spite M. 1982, Astr. Ap., **115**, 357.

Vauclair, S. 1987 in IAU Symposium 132, eds. G. Cayrel de Stroble & M. Spite, pg. 432.

Vauclair, S. 1988, in *Dark Matter*, eds. J. Audouze & J. Tran Thanh Vhan, pg. 269.

Vittorio N. 1993 in *CBR Workshop* in Santander , Spain

Zeldovich Ya.B., Kurt V.G. & Sunyaev R.A. 1968 Zh. Eksp. Teor. Fiz., **55**, 278

Yang, J., Turner, M.S., Steigman, G., Schramm, D.N., Olive, K. 1984 Ap.J., **281**, 493

New Constraints on Reionization from the Compton y-parameter

Max Tegmark[1] and Joseph Silk[2]

[1] Department of Physics, University of California, Berkeley, CA 94720
[2] Departments of Astronomy and Physics, and Center for Particle Astrophysics, University of California, Berkeley, CA 94720

1 Abstract

The most plausible cosmological model dominated by baryonic dark matter, characterized by primeval isocurvature fluctuations, may be ruled out in the limit of very early reionization by the new COBE FIRAS limit $y < 2.5 \times 10^{-5}$ on Compton y-distortions of the cosmic microwave background. A neutral phase is inevitable for such a model to survive this confrontation with data. Indeed, not only this model, but virtually any model with similar cosmological parameters ($\Omega_0 = \Omega_b \approx 0.15$, $\Omega_{igm} \approx 0.04$, $h \approx 0.8$), in which the intergalactic medium is ionized with temperatures above 10^4K as early as $z = 800$ is ruled out. This conclusion is independent of whether the main ionization mechanism is collisional ionization due to high temperatures or photoionization at lower temperatures. Completely ionized scenarios with typical CDM parameters ($\Omega = 1, h^2\Omega_b \approx 0.015$) are still allowed for most realistic spectra of ionizing radiation, but will also be ruled out if the y-limit were to be improved by another factor of a few. If the ionizing radiation has a typical quasar spectrum, then the y-constraint implies roughly $h^{3/2}\Omega_b\Omega_0^{-1/4} < 0.025$ for fully ionized models.

2 Introduction

Recombination of the primeval plasma is commonly assumed but was by no means inevitable. Theories exist that predict early reionization are as diverse as those invoking primordial seed fluctuations that underwent early collapse and generated sources of ionizing radiation, and models involving decaying or annihilating particles. The former class includes cosmic strings and textures, as well as primordial isocurvature baryon fluctuations. The latter category includes baryon symmetric cosmologies as well as decaying exotic paricles or neutrinos.

The Compton y-distortion of the cosmic microwave background (CMB) provides a unique constraint on the epoch of reionization. In view of the extremely sensitive recent FIRAS limit of $y < 2.5 \times 10^{-5}$, we have reinvestigated constraints on the early ionization history of the intergalactic medium (IGM), and have chosen to focus on what we regard as the most important of the non-standard recombination history models, namely the primordial isocurvature baryon scenario involving a universe dominated by baryonic dark matter (BDM), as advocated by Peebles (1987), Gnedin & Ostriker (1992) (hereafter "GO"), Cen, Ostriker & Peebles (1993) and others. This class of models takes the simplest matter content for the universe, namely baryons, to constitute dark matter in an amount that is directly observed and is even within the bounds of primordial nucleosynthesis, if interpreted liberally, and can reconstruct essentially all of the observed phenomena that constrain large-scale structure. The BDM model is a non-starter unless the IGM underwent very early reionization, in order to avoid producing excessive CMB fluctuations on degree scales. Fortunately, early nonlinearity is inevitable with BDM initial conditions, $\delta\rho/\rho \propto M^{-5/12}$, corresponding to a power-spectrum $\langle \delta_k^2 \rangle \propto k^{-1/2}$ for the observationally preferred choice of spectral index (Cen, Ostriker & Peebles 1993).

Is it possible that the IGM has been highly ionized since close to the standard recombination epoch at $z \approx 1100$? Perhaps the most carefully studied BDM scenario in which this happens is that by GO. In their scenario, $\Omega_0 = \Omega_{b0} \approx 0.15$. Shortly after recombination, a large fraction of the mass condenses into faint stars or massive black holes, releasing energy that reionizes the universe and heats it to $T > 10000K$ by $z = 800$, so Compton scattering off of hot electrons causes strong spectral distortions in the cosmic microwave background. The models in GO give a Compton y-parameter between 0.96×10^{-4} and 3.1×10^{-4}, and are thus all ruled out by the most recent observational constraint from the COBE FIRAS experiment, $y < 2.5 \times 10^{-5}$ (Mather et al. 1993).

There are essentially four mechanisms that can heat the IGM sufficiently to produce Compton y-distortions:

- Photoionization heating from UV photons (Shapiro & Giroux 1987, Donahue & Shull 1991)

- Compton heating from UV photons

- Mechanical heating from supernova-driven winds (Schwartz et al. 1975, Ikeuchi 1981, Ostriker & Cowie 1981)

- Cosmic ray heating (Ginzburg & Ozernoi 1965)

The second effect tends to drive the IGM temperature towards two-thirds of the temperature of the ionizing radiation, whereas the first effect tends to drive the temperature towards a lower value T^* that will be defined below. The third and fourth effect can produce much higher temperatures, often in the millions of degrees. The higher the temperature, the greater the y-distortion.

In the GO models, the second effect dominates, which is why they fail so badly. In this paper, we wish to place limits that are virtually impossible to evade. Thus we will use the most cautions assumptions possible, and neglect the latter three heating mechanisms altogether.

3 The Compton y-Parameter

Thomson scattering between CMB photons and hot electrons affects the spectrum of the CMB. It has long been known that hot ionized IGM causes spectral distortions to the CMB, known as the Sunyayev-Zel'dovich effect. A useful measure of this distortion is the Comptonization y-parameter (Kompanéets 1957, Sunyayev ans Zel'dovich 1969, Stebbins & Silk 1986, Bartlett & Stebbins 1991)

$$y = \int \left(\frac{kT_e - kT_\gamma}{m_e c^2} \right) n_e \sigma_t c \, dt = y^* \int \frac{(1+z)}{\sqrt{1+\Omega_0 z}} \Delta T_4(z) x(z) dz, \qquad (1)$$

where

$$y^* \equiv \left[1 - \left(1 - \frac{x_{He}}{4x} \right) Y \right] \left(\frac{k \times 10^4 K}{m_e c^2} \right) \left(\frac{3 H_0 \Omega_{igm} \sigma_t c}{8 \pi G m_p} \right) \approx 9.576 \times 10^8 h \Omega_{igm}.$$

Here T_e is the electron temperature, T_γ is the CMB temperature, $\Delta T_4 \equiv (T_e - T_\gamma)/10^4 K$, Ω_{igm} is the mass in intergalactic medium and $x(z)$ is the fraction of the hydrogen that is ionized at redshift z. Ω_{igm} is the fraction of the critical density in intergalactic hydrogen and helium. The integral is to be taken from the reionization epoch to today. In estimating the electron density n_e, we have taken the mass fraction of helium to be $Y \approx 24\%$ and assumed $x_{He} \approx x$, i.e. that helium never becomes doubly ionized and that the fraction that is singly ionized equals the fraction of hydrogen that is ionized. The latter is a very crude approximation, but makes a difference of only 6%.

Let us estimate this integral by making the approximation that the IGM is cold and neutral until a redshift z_{ion}, at which it suddenly becomes ionized, and after which it remains completely ionized with a constant temperature T_e. Then for $z_{ion} \gg 1$ and $T_e \gg z_{ion} \times 2.7$K we obtain

$$y \approx 6.4 \times 10^{-8} h \Omega_{igm} \Omega_0^{-1/2} T_4 \, z_{ion}^{3/2},$$

where $T_4 \equiv T_e/10^4 K$. Substituting the most recent observational constraint from the COBE FIRAS experiment, $y < 2.5 \times 10^{-5}$ (Mather et al. 1993), into this expression yields

$$z_{ion} < 554 T_4^{-2/3} \Omega_0^{1/3} \left(\frac{h \Omega_{igm}}{0.03} \right)^{-2/3}. \qquad (2)$$

Thus the only way to have z_{ion} as high as 1100 is to have temperatures considerably below 10000K. In the following section, we will see to what extent this is possible.

4 IGM Evolution in the Strong UV Flux Limit

In this section, we will calculate the thermal evolution of IGM for which
- the IGM remains almost completely ionized at all times,
- the Compton y-distortion is minimized given this constraint.

4.1 The Ionization Fraction

In a homogeneous IGM at temperature T exposed to a density of ζ UV photons of energy $h\nu > 13.6\,\mathrm{eV}$ per proton, the ionization fraction x evolves as follows:

$$\frac{dx}{d(-z)} = \frac{1+z}{\sqrt{1+\Omega_0 z}} \left[\lambda_{pi}(1-x) + \lambda_{ci} x(1-x) - \lambda_{rec} x^2 \right], \tag{3}$$

where $H_0^{-1}(1+z)^{-3}$ times the rates per baryon for photoionization, collisional ionization and recombination are given by

$$\begin{cases} \lambda_{pi} \approx 1.04 \times 10^{12} \left[h\Omega_{igm}\sigma_{18} \right] \zeta, \\[2mm] \lambda_{ci} \approx 2.03 \times 10^4 h\Omega_{igm} T_4^{1/2} e^{-15.4/T_4}, \\[2mm] \lambda_{rec} \approx 0.7173 h\Omega_{igm} T_4^{-1/2} \left[1.808 - 0.5 \ln T_4 + 0.187 T_4^{1/3} \right], \end{cases} \tag{4}$$

and $T_4 \equiv T_e/10^4 \mathrm{K}$. Here σ_{18} is the spectrally-averaged photoionization cross section in units of $10^{-18}\mathrm{cm}^2$. The differential cross section is given by (Osterbrock, 1974)

$$\frac{d\sigma_{18}}{d\nu}(\nu) \approx \begin{cases} 0 & \text{if } \nu < 13.6\,\mathrm{eV}, \\[2mm] 6.30 \frac{e^{4-4\,\mathrm{arctan}(\epsilon)/\epsilon}}{\nu^4 \left(1 - e^{-2\pi/\epsilon} \right)} & \text{if } \nu \geq 13.6\,\mathrm{eV}, \end{cases} \tag{5}$$

where

$$\epsilon \equiv \sqrt{\frac{h\nu}{13.6\,\mathrm{eV}} - 1}.$$

The recombination rate is the total to all hydrogenic levels (Seaton 1959, Spitzer 1968). Recombinations directly to the ground state should be included here, since as will become evident below, the resulting UV photons are outnumbered by the UV photons that keep the IGM photoionized in the first place, and thus can be neglected when determining the equilibrium temperature.

At high redshifts, the ionization and recombination rates greatly exceed the expansion rate of the universe, and the ionization level quickly adjusts to a quasi-static equilibrium value for which the expression in square brackets in equation (3) vanishes. In the absence of photoionization, an ionization fraction x close to unity requires $T_e > 15000\mathrm{K}$. Substituting this into equation (2) gives consistency with $z_{ion} > 1000$ only if $h\Omega_{igm} < 0.008$, a value clearly inconsistent with the standard nucleosynthesis constraints (Smith *et al.* 1993). Thus any reheating scenario that relies on collisional ionization to keep the IGM ionized at all times may be considered ruled out by the COBE FIRAS data.

However, this does not rule out all ionized universe scenarios, since photoionization can achieve the same ionization history while causing a much smaller y-distortion. The lowest temperatures (and hence the smallest y-distortions) compatible with high ionization will be obtained when the ionizing flux is so strong that $\lambda_{pi} \gg \lambda_{ci}$.

In this limit, to a good approximation, equation (3) can be replaced by the following simple model for the IGM:

- •It is completely ionized ($x = 1$).
- •When a neutral hydrogen atom is formed through recombination, it is instantly photoionized again.

Thus the only unknown parameter is the IGM temperature T_e, which determines the recombination rate, which in turn equals the photoionization rate and thus determines the rate of heating.

4.2 The Spectral Parameter T^*

The net effect of a recombination and subsequent photoionization is to remove the kinetic energy $\frac{3}{2}kT$ from the plasma and replace it with the kinetic energy $\frac{3}{2}kT^*$, where T^* is defined by $\frac{3}{2}kT^* \equiv \langle E_{uv} \rangle - 13.6\,\text{eV}$ and $\langle E_{uv} \rangle$ is the average energy of the ionizing photons. Thus the higher the recombination rate, the faster this effect will tend to push the temperature towards T^*.

The average energy of the ionizing photons is given by the spectrum $P(\nu)$ as $\langle E_{uv} \rangle = h\langle \nu \rangle$, where

$$\langle \nu \rangle = \frac{\int_0^\infty P(\nu)\sigma(\nu)d\nu}{\int_0^\infty \nu^{-1}P(\nu)\sigma(\nu)d\nu}.$$

Here σ is given by equation (5). Note that, in contrast to certain nebula calculations where all photons get absorbed sooner or later, the spectrum should be weighted by the photoionization cross section. This is because most photons never get absorbed, and all that is relevant is the energy distribution of those photons that do. Also note that $P(\nu)$ is the energy distribution (W/Hz), not the number distribution which is proportional to $P(\nu)/\nu$.

Table 1. Spectral parameters

UV	source	Spectrum	$P(\nu)$	$\langle E_{uv} \rangle$	T^*
O3	star	$T = 50000$K	Planck	17.3 eV	28300K
O6	star	$T = 40000$K	Planck	16.6 eV	23400K
O9	star	$T = 30000$K	Planck	15.9 eV	18000K
Pop. III star		$T = 50000$K	Vacca	18.4 eV	36900K
QSO		$\alpha = 1$ power law		18.4 eV	37400K
		$\alpha = 2$ power law		17.2 eV	27800K
		$\alpha = 0$ power law		20.9 eV	56300K
		$T = 100000$K	Planck	19.9 eV	49000K

The spectral parameters $\langle E_{uv} \rangle$ and T^* are given in Table 1 for some selected spectra. A power law spectrum $P(\nu) \propto \nu^{-\alpha}$ with $\alpha = 1$ fits observed QSO spectra rather well in the vicinity of the Lyman limit (Cheney & Rowan-Robinson 1981, O'Brien et al. 1988), and is also consistent with the standard model for black hole accretion. A Planck spectrum $P(\nu) \propto \nu^3 / (e^{h\nu/kT} - 1)$ gives a decent prediction of T^* for stars with surface temperatures below 30000K. For very hot stars, more realistic spectra (Vacca 1993) fall off much slower above the Lyman limit, thus giving higher values of T^*. As seen in Table 1, an extremely metal poor star of surface temperature 50000K gives roughly the same T^* as QSO radiation. The only stars that are likely to be relevant to early photoionization scenarios are extremely hot and short-lived ones, since the universe is less than a million years old at $z = 1000$, and fainter stars would be unable to inject enough energy in so short a time. Conceivably, less massive stars could play a dominant role later on, thus lowering T^*. However, since they radiate such a small fraction of their energy above the Lyman limit, very large numbers would be needed, which could be difficult to reconcile with the absence of observations of Population III stars today. The last three spectra in Table 1 do not correspond to any known types of objects, and have been added merely to illustrate that T^* is relatively insensitive to the actual shape of the spectrum.

4.3 The Thermal Evolution

At the low temperatures involved, the two dominant cooling effects are Compton drag against the microwave background photons and cooling due to the adiabatic expansion of the universe. Combining these effects, we obtain the following equation for the thermal evolution of the IGM:

$$\frac{dT}{d(-z)} = -\frac{2}{1+z}T + \frac{1+z}{\sqrt{1+\Omega_0 z}}\left[\lambda_{comp}(T_\gamma - T) + \lambda_{rec}(T)(T^* - T)\right], \qquad (6)$$

where

$$\lambda_{comp} = \frac{\pi^2}{45}\left(\frac{kT_\gamma}{\hbar c}\right)^4 \frac{\hbar\sigma t}{H_0 m_e}(1+z)^{-3} \approx 0.00104h^{-1}(1+z)$$

is $(1+z)^{-3}$ times the Compton cooling rate per Hubble time and $T_\gamma = T_{\gamma 0}(1+z)$. We have taken $T_{\gamma 0} \approx 2.726$K (Mather *et al.* 1993). Numerical solutions to this equation are shown in Figure 1, and the resulting y-parameters are given in Table 2.

Table 2. Compton y-parameters for various scenarios

Model		Ω_0	Ω_{igm}	h	T^*	z_{ion}	$y/2.5 \times 10^{-5}$	FIRAS verdict	
QSO	BDM I	0.15	0.04	0.8	37400K	1100	2.71	Ruled	out
QSO	BDM II	0.15	0.04	0.8	37400K	500	1.38	Ruled	out
QSO	BDM III	0.15	0.03	0.8	37400K	800	1.33	Ruled	out
O9	BDM	0.15	0.04	0.8	18000K	800	1.23	Ruled	out
QSO	CDM I	1	0.06	0.5	37400K	1100	0.71	OK	
QSO	CDM II	1	0.03	0.8	37400K	1100	0.66	OK	
Hot	CDM	1	0.06	0.5	49000K	1100	0.87	OK	

The temperature evolution separates into two distinct phases. In the first phase, which is almost instantaneous due to the high recombination rates at low temperatures, T rises very rapidly, up to a quasi-equilibrium temperature slightly above the temperature of the microwave background photons. After this, in the second phase, T changes only slowly, and is approximately given by setting the expression in square brackets in equation (6) equal to zero. This quasi-equilibrium temperature is typically much lower than T^*, since Compton cooling is so efficient at the high redshifts involved, and is given by

$$\Delta T \equiv T_e - T_\gamma \approx \frac{\lambda_{rec}}{\lambda_{comp}}(T^* - T_e) \approx \frac{690}{1+z}g(T_e)h^2\Omega_{igm}(T^* - T),$$

independent of Ω_0, where $g(T_e)$ encompasses the rather weak temperature dependence of λ_{rec}. Thus ΔT is roughly proportional to $h^2\Omega_{igm}T^*/z$ for $z \gg 1$. Substituting this into equation (1) gives the approximation

$$y \approx 0.01h^3\Omega_{igm}^2\Omega_0^{-1/2}T_4^*(z_{ion}/1100)^{0.8}, \tag{7}$$

which is accurate to about 10% within the parameter range of cosmological interest. (The exponent 0.8 fits the numerical data better than an exponent of 0.5, because of the behavior at low redshifts.) We have used equation (7) in Figure 2 by setting $y = 2.5 \times 10^{-5}$ and $z_{ion} = 1100$. The shaded region of parameter space is thus ruled out by the COBE FIRAS experiment for fully ionized scenarios.

5 Conclusions

The most plausible cosmological model dominated by baryonic dark matter, characterized by primeval isocurvature fluctuations, may be ruled out in the limit of very early reionization by the new COBE FIRAS limit $y < 2.5 \times 10^{-5}$ on Compton y-distortions of the cosmic microwave background. A neutral phase is inevitable for such a model to survive this confrontation with data. Indeed, not only this model, but virtually any model with similar cosmological parameters ($\Omega_0 = \Omega_b \approx 0.15$, $\Omega_{igm} \approx 0.04$, $h \approx 0.8$), in which the intergalactic medium is ionized with temperatures above 10^4K as early as $z = 800$ is ruled out. This conclusion is independent of whether the main ionization mechanism is collisional ionization due to high temperatures or photoionization at lower temperatures. Completely ionized scenarios with typical CDM parameters ($\Omega = 1$, $h^2\Omega_b \approx 0.015$) are still allowed for most realistic spectra of ionizing radiation, but will also be ruled out if the y-limit were to be improved by another factor of a few. If the ionizing radiation has a typical quasar spectrum, then the y-constraint implies roughly $h^{3/2}\Omega_b\Omega_0^{-1/4} < 0.025$ for fully ionized models. is possible only if most of the baryons form BDM when reionization occurs, and are thereby removed as a source of y-distortion, at least in the diffuse phase. This must be an unlikely occurrence at $z > 100$, since once the matter is reionized at this high a redshift, Compton drag is extremely effective in inhibiting any further gas collapse until $z < 100$. Since it takes only a small fraction of the baryons in the universe to provide a source of photons sufficient to maintain a fully ionized IGM even at $z \sim 1000$, we suspect that most of the baryons remain diffuse until Compton drag eventually becomes ineffective. Moreover, the possibility that the IGM is only partially reionized at $z \sim 1000$ (*e.g.* GO), a situation which allows a lower value of the y-parameter, seems to us to be implausible as a delicate adjustment of ionization and recombination time-scales over a considerable range in z would be required. A complementary argument that greatly restricts the parameter space allowable for fully ionized BDM models appeals to temperature fluctuations induced on the secondary last scattering surface, both by first order Doppler terms on degree scales and by second order terms on subarcminute scales (Hu, Scott and Silk 1993). Thus, BDM models would seem to be in serious difficulty because of the low limit on a possible y-distortion, with a window remaining only if late reionization occurs. However, this would mean that primary Doppler temperature fluctuations become difficult to completely erase (Sugiyama 1993), and provide an observable source of fluctuations on degree scales.

Finally, we note that with a modest improvement in the y limit, one would be able to assert that a neutral phase was inevitable in *any* model that has the minimum baryon density required by primordial nucleosynthesis, and in particular, for inflationary (that is, $\Omega \approx 1$) models.

The authors would like to thank W. Hu and D. Scott for many useful comments, and W. Vacca for providing stellar spectra. This research has been supported in part by a grant from the NSF.

References

1. Arons, J. & Wingert, D. W. 1972, Ap. J., **177**, 1.
2. Bartlett, J. & Stebbins, A. 1991, Ap. J., **371**, 8.
3. Cen, R., Gnedin, N. Y., Koffmann, L. A., & Ostriker, J. P. 1992, Preprint
4. Cen, R., Ostriker, J. P. & Peebles, P. J. E 1993, preprint
5. Cheney, J. E. & Rowan-Robinson, M. 1981, MNRAS, **195**, 831.
6. Couchman, H. M. P. 1985, MNRAS, **214**, 137.
7. Couchman, H. M. P. & Rees, M. 1988, MNRAS, **221**, 53.
8. Dalgarno, A. & McCray, R. A. 1972, A. Rev. Astr. Astrophys, **10**, 375.
9. Donahue, M. & Shull, J. M. 1991, Ap. J., **383**, 511.
10. Feynman, R. P. 1939, Phys. Rev., **56**, 340.
11. Ginsburg, V. L. & Ozernoi, L. M. 1965, Astron. Zh., **42**, 943.
    (Engl. transl. 196 Sov. Astron. AJ, 9, 726)
12. Gnedin, N. Y. & Ostriker, J. P.1992, Ap. J., **400**, 1.
13. Hu, W., Scott, D. and Silk, J. 1993, preprint.
14. Ikeuchi, S. 1981, Publ. Astr. Soc. Jpn., **33**, 211.
15. Kompanéets, A. 1957, Soviet Phys. – JETP, **4**, 730.
16. Mather *et al.* 1993, preprint.
17. O'Brien P.T., Wilson, R & Gondhalekar, P. M 1988, MNRAS, **233**, 801.
18. Osterbrock, D. E. 1974, *Astrophysics of Gaseous Nebulae* (Freeman, San Francisco)
19. Ostriker, J. P. & Cowie, C. F. 1981, Ap. J., **243**, L127.

20. Peebles, P. J. E. 1987, Ap. J., **315**, L73.
21. Schwartz, J., Ostriker, J. P., & Yahil, A. 1975, Ap. J., **202**, 1.
22. Seaton, M. 1959, MNRAS, **119**, 84.
23. Shapiro, P. R. & Giroux, M. L. 1987, Ap. J., **321**, L107.
24. Smith, M. S., Kawano, L. H. & Malaney, R. A. 1993, Ap. J. S., **85**, 219.
25. Spitzer, L. 1968, *Diffuse Matter in Space* (Wiley, New York).
26. Sugiyama, N. 1993, private communication.
27. Stebbins, A., & Silk, J. 1986, Ap. J., **300**, 1.
28. W. Vacca 1993, private communication.
29. Zel'dovich, Y., & Sunyaev, R. 1969, Ap. Space Sci., **4**, 301.

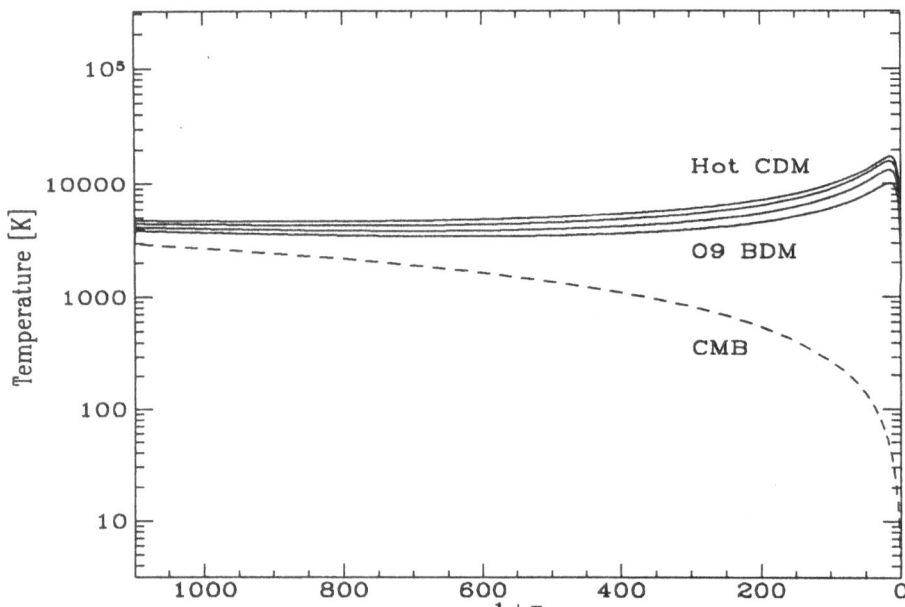

Figure 1. Thermal histories for various models

The temperature of the photoionized IGM is plotted for four of the cosmological models and spectra of ionizing radiation listed in Table 2. From top to bottom, they are Hot CDM, QSO BDM, QSO CDM and O9 BDM. The lowermost curve gives the temperature of the CMB photons.

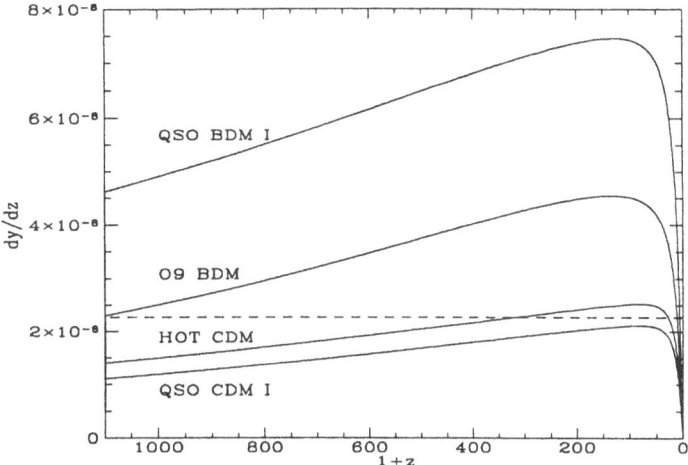

Figure 2. dy/dz **for various models**

The contribution to the y-parameter from different redshifts is plotted four of the cosmological models and spectra of ionizing radiation listed in Table 2. Thus for each model, the area under the curve is the predicted y-parameter. The area under the horizontal dashed line is 2.5×10^{-5}, i.e. the COBE FIRAS limit.

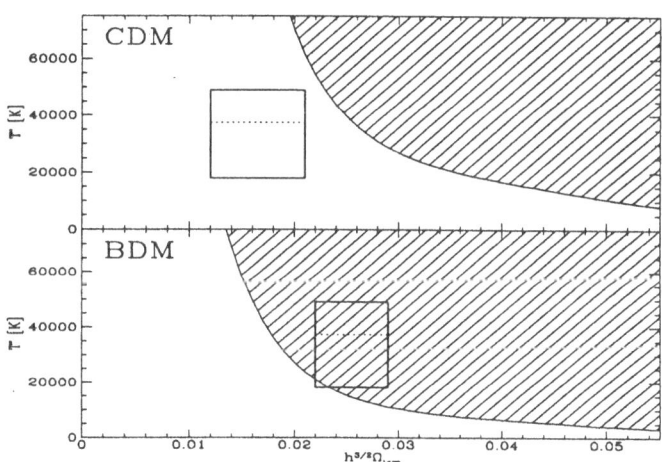

Figure 3. Predicted and ruled out regions of parameter space

The hatched regions of parameter space are ruled out by the the COBE FIRAS limit $y < 2.5 \times 10^{-5}$ for $z_{ion} = 1100$. $\Omega_0 = 1$ in the CDM plot and $\Omega_0 = 0.15$ in the BDM plot. The rectangular regions are the assumed parameter values for CDM and the GO BDM model, respectively. For CDM, the range $0.012 < h^{3/2}\Omega_{igm} < 0.021$ is given by the nucleosynthesis constraint $0.011 < h^2\Omega_b < 0.015$ and the assumption that $0.5 < h < 0.8$. (If $\Omega_{igm} < \Omega_b$, the rectangle shifts to the left.) For GO BDM, $h = 0.8$ and $0.03 < \Omega_{igm} < 0.04$. The vertical range corresponds to feasible values of the spectral parameter T^*. The upper limit corresponds to higly speculative star with surface temperature $100000K$ and $T^* = 49000K$. The lower line corresponds to an O9 star. The dotted horizontal limit corresponds to the spectrum expected from quasars/accreting black holes.

FUTURE PROJECTS ON THE COSMIC MICROWAVE BACKGROUND

Monique SIGNORE [1], Bianca MELCHIORRI [2] and Francesco MELCHIORRI[2]

1. Ecole Normale Supérieure, 24 rue Lhomond, Paris , France

2. La Sapienza, 2 P.le Aldo Moro, Roma, Italy

Abstract : The Cosmic Microwave Background (CMB) is consistent with a Planck Spectrum from 50 to 0.05 cm wavelengths. The importance of detailed measurements of the frequency and angular distributions of the CMB is briefly discussed. Future projects on the CMB are presented.

Keywords : cosmic microwave background

1. On spectral measurements of the CMB

The spectrum of the CMB provides us with a unique probe of the thermal history of the universe. Accordingly, measurements of, or upper limits on distortions can be used to constrain physical processes which could cause an energy release in the early universe. With the recent and precise determination of the CMB spectrum from 0.5 to 5mm given by the Far Infrared Absolute Spectrophotometer (FIRAS) on the COBE satellite, we ask the following questions : How to improve it ? In particular at larger wavelenghts than FIRAS ? But, first : what can we hope to learn by precisely measuring the CMB spectrum at short and long wavelengths ?

In the early universe, energy may have been released into the "primordial plasma" generating spectral CMB distortions. Several possibilities may occur depending on the epoch of energy injection : 1/ At $z > z_{max} \sim 8 \ 10^6$, double Compton emission $(e + \gamma -> e + \gamma + \gamma)$ acts efficiently and any cosmic event leaves no direct signature in the CMB spectrum ; 2/ For redshifts z such that : $z_{BE} < z < z_{max}$, any heat release generates a Bose-Einstein distortion characterized by a chemical potential μ ; the Bose-Einstein distribution is achieved as long as the universe is thick to Compton scattering $(\tau_c = 1$ at $z = z_{BE})$; 3/ At $z_r < z < z_{BE}$ the universe is no more optically thick to Compton scattering ($e + \gamma \longrightarrow e + \gamma$) characterized by the Compton parameter y ; the CMB spectrum becomes, then , a partially Comptonized spectrum. However, at these redshifts, free-free emission ($e+p \longrightarrow e + p + \gamma$) can also be significant ; it is characterized by its optical depth Y_{ff}. Therefore, Compton scaterring and free-free emission can compete until $z \sim z_r$, when free electrons combine with protons to form neutral hydrogen atoms. In standard cosmology, $z_r \sim$ 1100 , but the universe could become reionized at later epochs $(z_r << 1100)$. Moreover, some general features are to be given : a / Free-free emission creates new photons and therefore yields a rise in temperature at long wavelengths; b/ Compton-scatterring redistributes photons to shorter wavelengths and therefore yields a deficit at long wavelengths and therefore yields a deficit at long wavelengths and a sharp rise in temperature at short wavelengths.

Accordingly, spectral measurements of the CMB : i/ At short wavelengths ($300\mu m$ - 3mm; 1000 GHz - 100 GHz) are first sensitive to Compton or Bose-Einstein distortions ii/ At long wavelengths (3mm - 260cm ; 100 GHz - 0.4 GHz) are sensitive to Bose-Einstein and free-free distortions.

The shortwavelength measurements of FIRAS (5 to 0.5 mm) provide the following limits (Mather et al.1993) :

$| y | < 2.5 \ 10^{-5}$, $| \mu | < 3.3 \ 10^{-4}$ (95% CL)

Ground-based measurements at cm-wavelengths give the limits (Kogut,1993) :

$| \mu | < 7 \ 10^{-3}$, $| Y_{ff} | < 10^{-4}$

All these limits have interesting consequences for cosmology : they constrain many physical processes in the early universe (Wright et al. 1993). However, one remark is to be done : due to the difficulties of measuring the atmospheric and galactic foregrounds, long wavelength measurements are less precise than the measurements done by FIRAS. Significant improvement in the precision of long wavelenght experiments will require the use of balloons or satellites - see for instance, the project CRATE -.

Finaly, two examples which show that CMB spectrum and anisotropy measurements complement each other, are to be given :

Example 1 : We have already mentioned that the universe might become reionized at a late epoch $(z_r \ll 1)$. Such a late reionization is generally studied through the presence or the absence of anisotropies at given angular scales and for given frequencies (see section 2). But it could also be observed through Compton scattering and free-free emission. FIRAS limit on y strongly constrains a hot IGM but cannot rule out a IGM with the Te ~ Tr . Precise measurements of the cm-wavelength CMB spectrum could provide a free-free signature of a "late-reionization" of the universe. A spatial mission-like CRATE- will be able to provide it!

Example 2 : Absolute radiometry could detect CMB distortions ΔI defined by : $I_{tot} = I_{BB} + \Delta I$. Differential radiometry could detect the corresponding dipole anisotropy DI_d defined by : $DI_{tot} = DI_{BB} + DI_d$. Besides the absolute values ΔI and DI_d , one can introduce the brightness contrast : $\Delta I / I_{tot}$ and the dipole contrast : DI_d / DI_{tot} .

Differential radiometry can be the best strategy for detecting a distortion if the ratio : $R = (DI_d / DI_{tot}) \times (\Delta I / I_{tot})^{-1}$ is greater than 1. In particular : i/ for Compton distortion, R is greater than 1 at low frequencies $(v < 2 \text{ cm}^{-1}$ or 60 GHz ; $\lambda > 1.5$ cm) ii/ for Bose-Einstein distortion, R is always less than 1.

2.- On anisotropy measurements of the CMB

Temperature anisotropies in the CMB are a direct probe of density perturbations at z ~ 1000. Accordingly, measurements of, or upper limits on, anisotropies can be used to constrain structure evolution models. With the recent detection of temperature anisotropies by Differential Microwave Radiometer (DMR) on the COBE satellite, an important question arises : What is the best anisotropy measurement ? Or what range of i/ frequencies v , ii/ angular scales θ , are needed to understand the origin and evolution of cosmic structure ?

2.1. Two frequency bands for the CMB

There are important astronomical and technical differences above and below 90 GHz.

2.1.1 $v \leq 90$ GHz (≤ 3 cm^{-1}) ; $\lambda \geq 3;3$ mm

• The galactic continuum emission is the main foreground. It is a superposition of synchrotron radiation, produced by cosmic ray electrons moving through the interstellar magnetic field and free-free emission produced by very low energy electrons. These emissions are mapped at 408 MHz, 1.4 GHz with a poor accuracy. Moreover, an extrapolation of these maps at frequencies in the range of 30 to 90 GHz is needed. Therefore, future anisotropy measurements require new accurate maps of the galactic continuum .

• At these frequencies, a new technology of transistor amplifiers is replacing SIS (Superconductor-Insulator-Superconductor) amplifiers and Maser amplifiers. These are HEMTs (High Electron Mobility Transistors).

2.1.2- $v \geq 90$ GHz (3 cm^{-1}) ; $\lambda \leq 3.3$ mm

• The interstellar dust emission (ISD) is the main source of confusion. This emission is well mapped by IRAS at 100 μm and by DIRBE on board COBE at 100,150 and 240 μm.

• In this frequency range, two complementary types of receivers must be used :

   - heterodyne receivers provide very high spectral resolution - see P.Encrenaz, 1991.

   - for continuum receivers, bolometric detectors cooled at temperatures of the order of 100 mK must be used either with an Adiabatic Demagnetization Refrigerator (ADR) or with a Dilution-type Refrigerator.

2.2. Angular scales for CMB anisotropy measurements

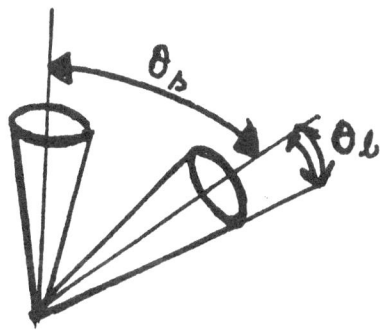

First, let us introduce θ_S and θ_b :

a typical double beam experiment measures temperature correlation over some separation angle θ_S , smoothed over an effective beam size θ_b

There are several different contributions to the amplitude of the CMB anisotropy ΔT/T that many authors have reviewed in these proceedings. Let us only recall that :

i/ on large angular scales (θ ≥ 5°) anisotropies are generated by spatial fluctuations in the curvature (Sach-Wolfe effect) and by fluctuations in the entropy (isocurvature effect). These are insensitive to the recombination history.

ii/ at intermediate angular scales, the main contribution arises from the Doppler shift across fluctuations at last scattering surface

iii/ over smaller angular scales (≤ 1°) the main sources of fluctuations are due to the finite thickness of the surface of last scattering; they are , therefore, extremely sensitive to the recombination history .

In particular :

- For the standard Cold Dark Matter Model (CDM) most of the power for the temperature anisotropies is distributed between $l = 100$ and $l = 1000$ for a spherical harmonic decomposition of the sky.

- If the universe remains ionized (or becomes reionized) primary fluctuations are suppressed on small angular scales and the anisotropy distribution is shifted into the range $l = 0$ to $l = 100$ - see J.Silk, A.Blanchard in these proceedings.

- Primary anisotropies at intermediate angular scales can also be erased by resonant molecular scattering , via primordial molecules such as LiH and LiH$^+$ molecules, at least in the radio frequency range (ν < 90 GHz) . Secondary anisotropies can also be induced by same primordial molecules , at smaller angular scales (< 30") - see F.Melchiorri in these proceedings and therefore observations through ground based and spatial missions devoted to anisotropy studies can help to solve the litium problem (see Signore et al.1993).

A remark is to be done : the CMB temperature anisotropies given above suppose that fluctuations originating in an inflationary epoch, grow by gravitational instability into the structures seen today. An alternative model for the structure formation involves the production of topological defects - such as cosmic strings, textures, etc..- arising in symmetry breaking phase transitions If, for example, strings are responsible for CMB anisotropies, experiments with a beam size $\theta_b \leq 2'$ could detect their non- gaussian signatures (R.Moessner et al. 1993).

Anisotropies produced by textures are reviewed in these proceeedings by S.Torres.

3. Projects

A list of : i) current balloon projects is given in Table 3.1 ; ii/ long duration balloonings (LDB) is given in Table 3.2 : BOOMERANG - a Berkeley-Rome project - and TOP HAT.See Cheng in these proceedings for TOP HAT.

BOOMERANG - a Berkeley - Roma collaboration - is an Antartica balloon program. The first flight is expected in December 1995. Each flight will have a typical duration of several days, a sky coverage of about 1000 pixels of 20 arminutes ^{2}, a per pixel sensitivity of 10^{-6}.

A list of satellite projects iii/ with experiments at low frequencies is presented in Table 3.3 ; iV/ with experiments at higher frequencies is presented in Table 3.4.

Let us only notice that :

a/ RELIKT 2 and AELITA are russian projects (see also M.Sazhin in these proceedings)

b/ CRATE and FIRE are NASA proposals for Explorer-class space experiments

c/ COBRAS, SAMBA and TOUTATIS are ESA proposals for medium missions (M3): COBRAS is an extension of DMR/COBE but with HEMTs (see also R.Mandolesi in these proceedings) . SAMBA is an up-to-date version of AELITA but with bolometer arrays cooled down to 0.1 K using a Dilution-type Refrigerator ; Let us remark that an experiment like SAMBA is no more than an extention to satellite of the balloon program BOOMERANG : it differs from it just for the sky coverage.TOUTATIS - with a heterodyne spectrometer - could provide a spectral analysis of the seven hot spots (> 300mK) revealed by DMR/COBE and could search for primordial molecules in protoclouds at $z \sim 10\text{-}15$.

d/ Let us notice that each of the above ESA proposals is presented as a propaedeutic to the cornerstone project FIRST of the "Horizon 2000" plan of ESA.

Some remarks must be done on ground based experiments which are reviewed in these proceedings by R.Davis, A.Lasenby and R.Rebolo :

. the IRAM 30 m telescope (Pico Veleta, Spain) with a 12.5 arcsec FWHM beam at 1.3 mm is well suited for the search of LiH molecules at high redshifts (de Bernardis et al. 1993) but also for the search of secondary CMB anisotropies

. a 4 -channel photometer (0.85, 1.1, 1.4 , 2.2 mm) with bolometers cooled to 0.3K (NEP ~ 10-15 W /$\sqrt{}$ Hz) is under construction; this photometer which will be matched to the MITO telescope (Millimetric and Infrared Testagrigia Observatory, Italy) , will have a field of view of 17 arcminutes.

. the bolometer DIABOLO can be matched either to the IRAM telescope at 2mm with θ_b = 35", θ_s ~1' or to the MITO telescope with θ_b=8' and θ_s = 16' . DIABOLO and the PRONAOS balloon are dedicated to the SZ effect (Sunayev-Zeldovich) . This SZ effect is reviewed in these proceedings by M.Birkinshaw.

4. Conclusions

For CMB spectrum, precise and thus spatial measurements at long wavenlengths are needed; measurements of, or upper limits on, free-free distortions would then be clear-cut tests for a late reionization of the universe.

For CMB anisotropy, there are many measurements projects. One must emphasize their technological requirements : i/ at low frequencies, HEMTs must be used, ii/ at high frequencies, photometry should be carried out with bolometric detectors at T ~ 100mK (via ADR or Dilution-type Refrigerators). Their scientific objectives are dealing with the fundamental problem of the origin of the structures in the universe.

We acknowledge many people who have provided us with their projects. In particular , we would like to thank X.Désert, P.Encrenaz, P.Mauskopf, J.L. Puget and P.Richards. This work was supported by the European Economic Community i/ Twinning Project 90 1000 13/JU 1, "Anisotropies of the Cosmic Background " (SCI-0531-c) ii/ Human Capital Project CHRX-CT92-0079, "The CMB radiation : measurements and interpretation".

References

- de Bernardis P., Dubrovich V., Encrenaz P., Maoli R., Masi S., Mastrantonio G., Melchiorri B., Melchiorri F., Signore M., and Tanzilli P.E. 1993, A & A, 269, 1 .

- Encrenaz P. 1993 in "IVth Rome Meeting on Astrophysics Cosmology and Particle Physics" F.Melchiorri ed., in press.

- Kogut A., 1993, in Proceeding of the International School of Astrophysics "D.Chalonge", N.Sanchez edit., in press

- Mather J.C., 1993, COBE preprint 93-01, submitted to Ap.J.

- Moessner R., Perivolaropoulos L. and Brandenberger R., 1993, preprint Brown-HET-911.

- Signore M., Vedrenne G., de Bernardis P., Dubrovich V., Encrenaz P., Maoli R., Masi S., Mastrantonio G., Melchiorri B., Melchiorri F., and Tanzilli P.E., 1993, ApJS, in press.

-Wright J.C., 1993, COBE preprint 93-03, submitted to Ap.J.

N.B. : Let us only note that the Space Science Advisory Committee (SSAC) , at its meeting held on 23/24 September 1993 , recommanded to the ESA Executive :

- " An assessment study of a mission to investigate the cosmic microwave background based on the COBRAS proposal, but taking into account also the SAMBA proposal, with scientists from tboth tcams being brought together in oder to define a small European mission".

TABLE 3.1 : Balloons

balloons	angular scales (θ_b, θ_s)	wavelength range or frequency range	sensibility	sky coverage
ARGO	20', 1°	4 channels: 0.5 ; 0.8 ; 1; 2mm	NET ~ 0.7mK/$\sqrt{}$ Hz R ~ 30 μK/pixel $\Delta T/T \sim 10^{-5}$	5° x 60°
ULISSE	4°, 6°	0.4 - 0.8 mm 0.8 - 2 mm	NET ~ 0.5 mK/$\sqrt{}$Hz R ~ 30 μK/pixel $\Delta T/T \sim 10^{-5}$	30° x 80°
MAX	0.5°, 1.3°	3 channels: 0.8 ; 1.3 ; 2 mm	NET ~ 0.5 mK /$\sqrt{}$Hz R ~ 15μK/pixel $\Delta T/T \sim 5\ 10^{-6}$	2° x 20°
MSAMI	0.5°	0.4 to 2 mm	$\Delta T/T \sim 10^{-5}$	
MSAMII	0.5°	3 channels: 2 to 4.3 mm	$\Delta T/T \sim 10^{-6}$	
PRONAOS	3'5 - 7'	0.54 - 1.2 mm	$\Delta T/T \sim 6\ 10^{-5}$	

TABLE 3.2 : Long Duration Balloonings

L D B	angular scales (θ_b , θ_s)	wavelength range frequency range	sensibility	sky coverage
BOOMERANG	30' , 2°	4 channels : 0.8 ; 1 ; 2 ; 3 mm	NET ~ 0.3 mK/ √Hz R ~ 3 μK / pixel ΔT/T ~ 10^{-6}	40° x 60°
TOP HAT	0.5° , 3°	5 channels : 0.5 to 2mm	R ~ 1 μK/pixel ΔT/T ~ 3 10^{-7}	

Table 3.3 : Satellites; low frequency range

Satellites	angular scales	wavelength range or frequency range	sensibility
RELIKT 2	0.5°	1.35 ; 0.87 ; 0.5 ; 0.15 cm	NET~ 0.3mK/√ Hz
CRATE		7.5 ; 3.8; 2.5; 0.9 cm	R ~ 25μK/pixel ΔT/T ~ 8 10^{-6}
PSI	0.5°-1° to 10°	40 ; 85-115 ; 130-160GHz	ΔT/T ~ 10^{-6}
COBRAS	0.5°	31.5 ; 53 ; 90 ; 130 GHz	ΔT/T ~ 10^{-6}

Table 3.4 : Satellites : high frequency range

satellites	angular scales	wavength range or frequency range	sensibility
AELITA	10'	0.3 to 3 mm	ΔT/T ~ 10^{-6}
FIRE	7'	0.2 to 2 mm	R ~ 1μk/pixel ΔT/T ~ 3 10^{-5}
FIRP/IRTS	0.5°	4 channels : 0.15 ; 0.25 ; 0.4 ; 0.7 mm	
SAMBA	20' to few degres	4 channels : 0.4 to 2.5 mm	ΔT/T ~ 1.5 10^{-6}

The COBRAS Mission

N.Mandolesi[1], G.F.Smoot[2] and M.Bersanelli[3]

[1] ITESRE - CNR, 40126 Bologna, Italy
[2] Lawrence Berkeley Laboratory, Berkeley, CA 94720, USA
[3] IFCTR - CNR, 20133 Milano, Italy

ABSTRACT - The Cosmic Background Radiation Anisotropy Satellite (COBRAS) is intended to image significant portions of the sky at $\sim 0.5°$ angular resolution and $\Delta T/T \sim 10^{-6}$ sensitivity. The project has been submitted to the European Space Agency (ESA) as a small-medium size European mission. We outline the motivation and mission concept on the basis of preliminary studies.

1 Introduction

The recent detection of primordial CMB fluctuations by the COBE-DMR experiment (Smoot et al. 1992) has provided remarkable support for our present cosmological understanding, particularly to current predictions from gravitational instability and inflationary scenarios. Far from saturating interest in the field, this achievement has established the basis of new fundamental steps and stimulated greater activity. The $> 7°$ angular resolution of the DMR instrument is sensitive to primordial fluctuations but does not provide information on scales $\simeq 1°$, corresponding to the horizon size on the last scattering surface. Only accurately probing the CMB angular distribution at these angular scales over large areas of the sky will observationally characterize the density fluctuations which originated present large scale structures ($\lesssim 100$ Mpc).

A number of balloon- and ground-based anisotropy experiments have been recently proposed or are presently underway to test these angular scales. Some results have already been obtained (e.g. ACME: Meinhold & Lubin 1991, Gaier et al. 1992; ARGO: De Bernardis et al. 1990; MAX: Fisher et al. 1992, Alsop et al. 1992, Meinhold et al. 1993, Gundersen et al. 1993; MIT: Meyer et al. 1991, Ganga et al. 1993, MSAM: Cheng et al. 1993) and more are expected in the near future. While these efforts provide outstanding observations, even under optimistic extrapolation of receiver sensitivities and resources, they will not cover a substantial ($\sim 10\%$) fraction of the sky with the required sensitivity (few $\times 10^{-6}$), necessary to link information on fluctuations at small and large (DMR) angular scales.

The COBRAS (Cosmic Microwave Background Anisotropy Satellite) mission has been conceived to meet this major objective. A proposal was submitted by a large international collaboration in response to the *ESA M3 Call for Mission Ideas* on May 1993 (Mandolesi et al. 1993).

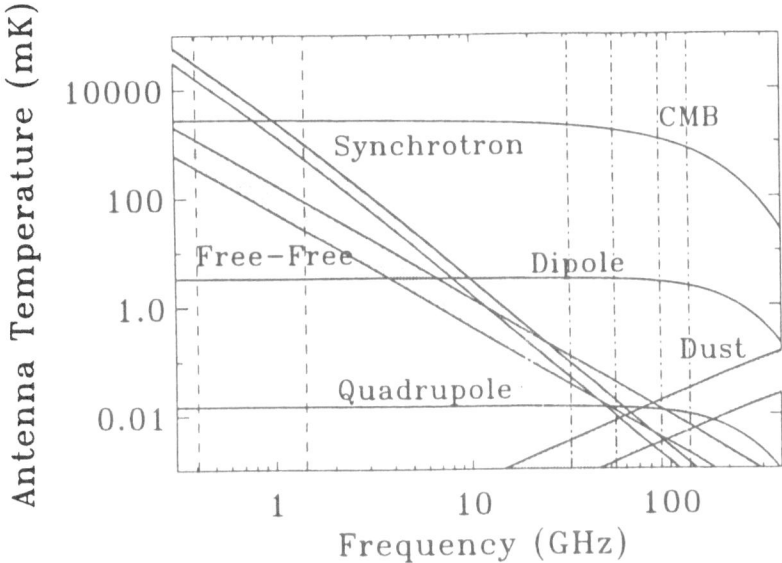

Fig. 1. Antenna temperature of Galactic foreground emission. Shown are the intensities of synchrotron emission, free-free emission, and thermal emission from interstellar dust. The width of the bands indicates the estimated variation in the range of Galactic latitudes $15° < |b| < 70°$. The vertical lines show the frequencies of existing Galactic maps (0. 408 GHz and 1.42 GHz) and the four COBRAS frequencies (31.5, 53, 90 and 125 GHz). Shown is also the antenna temperature of the CMB and the amplitude of the CMB dipole and quadrupole.

2 Requirements

The objective of COBRAS is to map 5 or 6 regions of $20° \times 30°$, or about 10% of the sky, with a sensitivity $\Delta T/T \sim 10^{-6}$ at an angular resolution of $\sim 0.5°$. Passive cooling is a key feature as such a goal requires a long lifetime (3–5 years) mission. The technical simplicity and extraordinary performance and rate of improvement of HEMT (High Electron Mobility Transistors) amplifiers (e.g. Mishra et al. 1988, Chao et al. 1990) strongly favour this technology for the COBRAS mapping instrument. Low noise HEMTs are already available at frequencies up to 50–60 GHz and prototypes are studied near 100 GHz. Passively cooled ($\sim$ 100 K) HEMTs with more reliable and space-qualified technology are soon expected to reach performances comparable to cryogenically-cooled bolometers.

The recent dramatic improvements in receiver noise push the new generation of CMB anisotropy experiments against the limit of confusing foreground emission. Multifrequency observations spanning a sufficiently wide spectral range, in conjunction with information from external surveys at frequencies where each foreground is dominant, are key features for reliably evaluating and removing foreground structures. We propose for COBRAS four frequencies in the range

$30 < \nu < 130$ GHz, i.e. in the cleanest window existing between synchrotron and free-free Galactic emission at low frequencies, and thermal emission from various dust components at higher frequencies (Fig. 1). The contribution of extragalactic radio sources also has a minimum in this spectral range (see e.g. Franceschini et al. 1989). We foresee the use of the same three frequencies as the COBE-DMR instrument to directly link COBRAS and DMR maps, and a higher frequency channel near 125 GHz. The covered spectral range will provide direct information on Galactic synchrotron and bremsstrahlung (relatively well known emission mechanisms) and minimize the contribution from dust emission, monitored by the 125 GHz channel. The highest and lowest frequency channels are sensitive guards against potential foreground signals. Table 1 summarizes the COBRAS main requirements.

Table 1
COBRAS Specifics and Requirements

Angular resolution	$\sim 0.5°$
Sensitivity	$\delta T/T \sim 10^{-6}$
	~ 1 mK/$\sqrt{s}$
Sky coverage	$\sim 10\%$
Frequencies	31.5, 53, 90, 125 GHz
Bandwidth	$\sim 10\%$
Optics	Off-axis Gregorian
	1.5 m primary (oversized)
Detection	Focal plane array
	Double-band feed horns
	HEMT radiometers ($T_{sys} < 100$ K)
	Passive cooling
Mission lifetime	~ 5 years
Orbit	LEO, Lagrangian Point 2, Heliocentric
Sidelobe rejection	-70 dB
	-80 dB (with ground screens)
Payload mass	~ 90 Kg
Payload power	~ 50 W
Data rate	~ 2 Kbit/sec

3 COBRAS Payload and Operation

We have considered a number of possible solutions for the optics of the instrument, including solutions involving microwave lenses or frequency sensitive surfaces. In conclusion, however, we opted for a more conventional off-axis Gregorian dual-reflector, similar in design to other instruments used for CMB anisotropy experiments. The primary reflector has an oversized aperture of ~ 1.5-m diameter and determines the overall size of the payload. An array of 26 dual-band feeds will be placed at the focus of the telescope, yielding 13 channels at each of the four frequencies. Preliminary simulations for this configuration show that coma lobes of the most decentered elements in the array can be contained below −50 dB level with acceptable projected beam symmetry. A schematic of the model payload is shown in Fig. 2, with the preliminary geometry of the array of focal plane feeds. The payload mass can be contained within 100 Kg without significant compromises of design or performance.

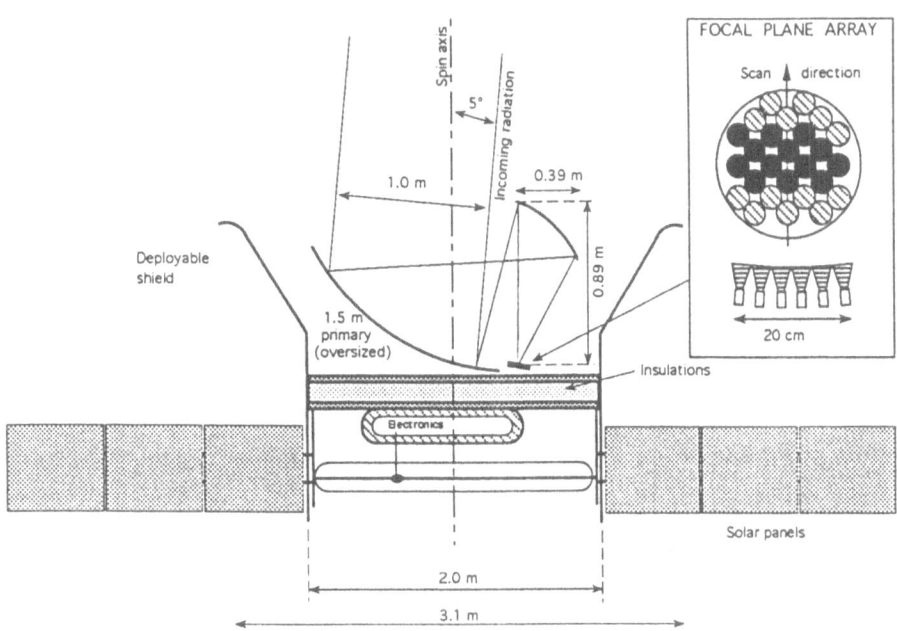

Fig. 2. Schematic COBRAS payload concept. The off-axis Gregorian telescope points at an offset angle ($\sim 5°$) from the spin axis. The inset shows a schematic of the preliminary configuration of the focal plane instrument. Dark circles represent 90–125 GHz dual band feeds; shaded circles 31.5–53 GHz feeds.

The high sensitivity hardens the requirements on potential systematic errors. Earth radiation needs to be rejected at the 10^{-8} level. With proper ground screen design appropriate earth emission rejection is feasible. Earth radiation and thermal performance considerations would favour far-Earth orbits such as Lagrangian Point 2 or Heliocentric. From such orbits the solid angle subtended by the Earth is a factor of 10^3 to 10^4 smaller than from a Low-Earth Orbit (LEO), strongly relaxing the sidelobe rejection requirements. However, we expect the choice of a LEO to be compatible with the COBRAS requirements with sufficiently careful design of the ground screen and pointing strategy. In the assessment phase we anticipate a detailed study of the antenna–reflectors–ground screen system and related performance in conjunction with pointing strategy and choice of orbit.

The proposed observational strategy is a circular scan (allowing spin-stabilization of the spacecraft) with periodic adjustments of the spin axis in a raster pattern. At COBRAS angular resolution only a fraction of the sky can be imaged to a sensitivity of 10^{-6} in a reasonable lifetime. The half of the sky within $\pm 30°$ Galactic latitude would not provide useful cosmological information since it is largely dominated by Galactic foregrounds. COBRAS will cover a ~ 600 square degrees region at $\Delta T/T \sim 10^{-6}$ in approximately one year. Five or six such sky patches will be selected for minimal power of Galactic and extragalactic confusion sources, and also based on updated results from the COBE-DMR analysis. COBRAS will perform interleaved observations of all sky patches according to an optimization pattern producing consistently improving maps of the selected regions over the mission lifetime.

Due to the long lifetime and flexibility of the mission, during the COBRAS assessment we plan to study the possibility of a lower sensitivity whole sky scan. This will provide information on the foreground and a cross comparison with the COBE DMR that would not be available otherwise.

4 Conclusions

There is a strong consensus in the scientific community that the determination of the CMB fluctuation spectrum from $\sim 0.5°$ up to and overlapping the angular scales probed by COBE-DMR is probably the most critical issue to be addressed by the next generation of cosmological observations. Only a long lifetime space mission can provide sufficient sensitivity and the extensive mapping capability to fully respond to this challenge. At the same time the development of a mission like COBRAS would certainly provide further motivation to future balloon and ground based experiments, as coordinated, intensive observations with extended spectral coverage of well defined sky areas will be necessary to further study the nature of the detected structures. COBRAS is designed to use relatively simple and space qualified technology, in many ways already successfully implemented in recent CMB observations.

ACKNOWLEDGEMENTS - We acknowledge the assistance of E.Pagana and P.Mantica from FPM-Space, Torino, in the study of the optical configuration and simulated performances. We acknowledge the support of Consiglio Nazionale delle Ricerche (CNR) and of LBL through DOE Contract DOE-AC-03-76SF0098.

References

Alsop, D.C., et al. 1992, ApJ, 395, 317

Chao, P.C., et al. 1990, Electr. Lett., 26, 27

Cheng, E.,S. et al. 1993, ApJ, submitted

De Bernardis, P. et al. 1990, ApJ, 360, L31

Fisher, M.L., Alsop, D.C., Cheng, E., Clapp, A., Cottingham, D., Gundersen, J., Koch, T., Kreisa, E., Meinhold, P., Lange, A., Lubin, P., Richards, P.L., Smoot, G.F. 1992, ApJ, 388, 242

Franceschini, A., Toffolatti, L., Danese, L., De Zotti, G. 1989, ApJ, 344, 35

Gaier, T., Shuster, J., Gundersen, J., Koch, T., Seiffert, M., Meinhold, P., Lubin, P. 1992, ApJ, 398, L1

Ganga, K. et al. 1993, ApJ, 410, L1

Gundersen, J.O., Clapp, A.C., Devlin, M., Holmes, W., Fisher, L., Meinhold, P., Lange, A., Lubin, P., Richards, P., Smoot, G. 1993, ApJ, 413, L1

Mandolesi, N., Smoot, G.F., Bersanelli, M., Cesarsky, C., Lachieze-Rey, M., Danese, L., Vittorio, N., De Bernardis, P., Dall'Oglio, G., Sironi, G., Crane, P., Janssen, M., Partridge, B., Beckman, J., Rebolo, R., Puget, J.L., Bussoletti, E., Raffelt, G., Davies, R., Encrenaz, P., Natale, V., Tofani, G., Merluzzi, P., Toffolatti, L., Scaramella, R., Martinez-Gonzales, E., Saez, D., Lasenby, A., and Efstathiou, G. 1993, *COBRAS Proposal*. Submitted to ESA M3 Call for Mission Ideas.

Meinhold, P. & Lubin, P. 1991, ApJ, 370, L11

Meinhold, P., Clapp, A.C., Devlin, M., Gundersen, J.O., Holmes, W., Fisher, L., Lange, A., Lubin, P., Richards, P., Smoot, G. 1993, ApJ, 409, L1-L4.

Meyer, S., Cheng, E.S., and Page, L. 1991 ApJ, 371, L7

Mishra, U.K. et al. 1988, IEEE Elec. Dev. Lett., 12, 647

Smoot, G.F. et al. 1992, ApJ 396, L1

Lecture Notes in Physics

For information about Vols. 1–388
please contact your bookseller or Springer-Verlag

New Series m: Monographs